HUBBARD OPERATORS IN THE THEORY OF STRONGLY CORRELATED ELECTRONS

HUBBARD OPERATORS IN THE THEORY OF STRONGLY CORRELATED ELECTRONS

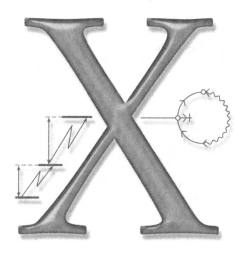

S.G. Ovchinnikov
V.V. Val'kov

L.V. Kirensky Institute of Physics,
Siberian Branch of the Russian Academy of Sciences
and Krasnoyarsk State Technical University, Russia

Imperial College Press

Published by

Imperial College Press
57 Shelton Street
Covent Garden
London WC2H 9HE

Distributed by

World Scientific Publishing Co. Pte. Ltd.
5 Toh Tuck Link, Singapore 596224
USA office: 27 Warren Street, Suite 401–402, Hackensack, NJ 07601
UK office: 57 Shelton Street, Covent Garden, London WC2H 9HE

British Library Cataloguing-in-Publication Data
A catalogue record for this book is available from the British Library.

The authors and publisher would like to thank the American Physical Society and Springer-Verlag for permission to reproduce selected materials in this book.

HUBBARD OPERATORS IN THE THEORY OF STRONGLY CORRELATED ELECTRONS

Copyright © 2004 by Imperial College Press

All rights reserved. This book, or parts thereof, may not be reproduced in any form or by any means, electronic or mechanical, including photocopying, recording or any information storage and retrieval system now known or to be invented, without written permission from the Publisher.

For photocopying of material in this volume, please pay a copying fee through the Copyright Clearance Center, Inc., 222 Rosewood Drive, Danvers, MA 01923, USA. In this case permission to photocopy is not required from the publisher.

ISBN-13 978-1-86094-430-7
ISBN-10 1-86094-430-2

Typeset by Stallion Press
Email: sales@stallionpress.com

Printed in Singapore

Contents

Preface   ix

PART 1

Chapter 1.   Hubbard Model as a Simplest Model of Strong
Electron Correlations   1

1.1   Hamiltonian of the Hubbard Model and its Symmetry . . .   1
1.2   Time Inversion   3
1.3   Electron Hole Symmetry   3
1.4   Pseudospin Symmetry for Half-Filled Case   4
1.5   The Band Limit   5
1.6   The Atomic Limit, Hubbard 1 Solution   6
1.7   The Hubbard Model in One and Infinite Dimension Cases .   9
1.8   The t–J Model   12
1.9   The Hubbard Model in X-Operator Representation   14
1.10 Study of the Electronic Structure by the Angle Resolved
Photoemission Spectroscopy   23

Chapter 2.   Multielectron Models in X-Operator Representation   27

2.1   The Isotropic Heisenberg Model   27
2.2   The Heisenberg Model with Single-Ion Anisotropy   29
2.3   The Atomic Representation for s–$d(f)$ Exchange Model . .   30
2.4   The Periodic Anderson Model   33
2.5   The Multiband Model of Transition Metal Oxides   37

Chapter 3. General Approach to the Quasiparticle
Description of Strongly Correlated Systems   41

3.1 The Definition of Fermi- and Bose-Type Quasiparticles . . .   41
3.2 The Exact Intracell Local Green's Function   44
3.3 The Generalized Tight-Binding Method for Quasiparticle
    Band Structure Calculations   46
3.4 The Concentration Dependence of the Quasiparticle
    Band Structure   52
3.5 *Ab Initio* Approach to the Quasiparticle Band
    Structure of Strongly Correlated Electron Systems   56
3.6 The Generalized Tight-Binding Method and the Exact
    Lehmann Representation   59

Chapter 4. Unitary Transformation Method in Atomic
Representation   63

4.1 Unitary Operators and Transformation Laws in Atomic
    Representation   63
4.2 Diagonalization of Two-Level and Quasi-Two-Level Forms .   67
4.3 The Diagonalization of Three-Level Forms   69
4.4 The Diagonalization of the N Level Hamiltonian   76

Chapter 5. Diagram Technique in the X-Operators
Representation   77

5.1 Green's Function in the X-Operators Representation   78
5.2 Wick's Theorem for the Hubbard Operators   82
5.3 Averaging the Diagonal X-Operators   86
5.4 External and Internal Operators: Simplest Vertexes   90
5.5 A Hierarchy of Interaction Lines and a
    Topological Continuity Principles   97
5.6 Calculation of the Sign for an Arbitrary Diagram   99
5.7 Diagrams with Ovals   104
5.8 General Rules for Arbitrary Order Diagram   112
5.9 The Larkin Equation   115
5.10 The Self-Energy and the Strength Operator   123
5.11 Low Temperature Properties of Anisotropic Ferromagnet . .   129
5.12 Effect of the Three-Site Correlated Hopping on the
    Magnetic Mechanism of $d_{x^2-y^2}$-Superconductivity
    in the $t-J^*$ Model   138

PART 2

Chapter 6. Spectral Properties of Anisotropic Metallic
Ferromagnets — 149

6.1 The Hamiltonian of the Anisotropic $s-f$ Magnet in Atomic Representation — 150
6.2 The Green Function and Dispersion Equation — 154
6.3 The Ferromagnetic Metal with Single Axis Anisotropy — 158
6.4 The Metallic Ferromagnets with the SA of Cubic Symmetry — 166

Chapter 7. Peculiarities of the de Haas-van Alphen Effect in Strongly Correlated Systems with Magnetic Polaron States — 175

7.1 The Hamiltonian of a Strong Correlated Narrow Band Antiferromagnet — 177
7.2 The Hamiltonian in the Non-colinear Phase: Unitary Transformation — 178
7.3 Construction of the Atomic Representation Basis — 179
7.4 The Hamiltonian of the Narrow Band Antiferromagnet in the Non-colinear Phase in Atomic Representation — 183
7.5 The Green's Functions and Dispersion Relations — 184
7.6 The Spectrum of Holes Near the Spin-flip Transition — 187
7.7 The Peculiarities of the de Haas-van Alphen Effect in Antiferromagnetic Semimetal with Magnetic Polaron States — 188
7.8 The Temperature Quantum Oscillations Caused by Non-Fermi Liquid Effects — 191

Chapter 8. The Electronic Structure of Copper Oxides in the Multiorbital p–d Model — 195

8.1 The Exact Diagonalization of the CuO_4 Cluster — 196
8.2 The Construction of the Wannier Functions and X-Representation of the Multiband p–d Model — 200

8.3 The Evolution of the Band Structure and ARPES Data with Doping from the Undoped Antiferromagnetic Insulator to Optimal Doped and Overdoped Metal 210

8.4 Comparison of the Electronic Structure and Superconductivity of Cuprates and Ruthenates 218

References 231

Index 239

Preface

Systems with strong electron correlations (SEC) have been known for a long time, since Mott (1949) explained the controversy in the insulating properties of NiO and Fe_3O_4, and the rule of one-electron band filling which results in the metal states of these oxides. The SEC effects determine the localization of electrons, thermodynamic and kinetic properties of a large group of d-, f-metals and their compounds. While a conventional single-electron band theory was successfully used to describe normal metals and alloys, its application to the SEC systems often yields incorrect results. That is why the problem of SEC is one of the essential problems of the modern condensed matter physics.

Recent discovery of new SEC systems such as the copper oxide-based high temperature superconductors and manganese oxides with colossal magnetoresistivity has raised the interest in the problem of SEC. Intensive studies all over the world have given insights to the ground state and elementary excitations in SEC systems. Very important results have been obtained by numerically exact computations like the quantum Monte-Carlo (QMC) and the exact diagonalization methods in the Hubbard model, t–J and p–d models for finite clusters (see e.g. reviews by Dagotto (1994); Kampf (1994); Brenig (1995); Ovchinnikov (1997)). Some progress achieved for the infinite systems is due to the development of diagram techniques for the Hubbard operators (Zaitsev 1975, 1976; Izyumov et al 1994), projection operators method (Fulde 1991) and the self-consistent treatment of the Hubbard model by a fluctuation exchange (FLEX) approach (Bickers et al 1989) and the dynamical mean-field theory in infinite dimensions (Metzner and Vollhardt 1989; Georges et al 1996).

Analysis of the Mendeleev periodic table shows that SEC-determined localization of electrons is typical for the $4f$-electrons of the rare-earth metals except Ce which is close to the localization–delocalization border. In the actinides, several elements in the beginning of the series (*Th, Pa, U, Np*, and *Pu*) are itinerant systems while Am and the remaining elements are localized electron systems. All $3d$, $4d$ and $5d$ elements are itinerant systems, nevertheless the SEC effects are known for $3d$ metals too. Along with electron localization, SEC also results in the correlation narrowing of the bandwidth and the dependence of the quasiparticle band structure on temperature, magnetic field and concentration. For example, a recent experimental study of $3d$ metals by X-ray magnetic circular dichroism of photoemission spectra has revealed SEC effects in *Ni* and *Co* (Schneider *et al* 1996; Ovchinnikov 1999). The decrease of kinetic energy due to reduced coordination number of the nearest neighbors in transition from bulk crystals to thin films and surfaces increases the role of SEC.

Now pass over from $3d$-elements to their compounds, SEC effects increase with distance between $3d$-nearest neighbors. It is especially apparent in oxides and other materials with dominant ionic bonding. For example, in $3d$-metal oxides the localization–delocalization border runs through *Ti* and *V* oxides. All $3d$-metal oxides following Cr are insulators with localized d-electrons. In $3d$-sulphides, the covalent bonding is more significant then in oxides and the border of electron localization is shifted to the middle of $3d$ group. In compounds, contribution of s, p, and d-orbital forms multiband energy structure. Some of these electrons may be itinerant while the other are localized. Near the localization border, the electrons are partly localized and partly itinerant. All these factors generate a large variety of electronic structures and physical properties of SEC systems. Examples are provided by copper and manganese oxides, heavy fermions, mixed valence systems and magnetic semiconductors.

The most popular models of SEC systems are the Hubbard model (Hubbard 1963), the $s-d(f)$ exchange model (Shubin and Vonsovskii 1934), the impurity Anderson model (Anderson 1961), the periodic Anderson model and the $p - d$ model of copper oxides (Emery 1987; Varma *et al* 1987; Gaididei and Loktev 1988). The main aim of this book is to give a description of the energy spectra and physical properties of the different SEC systems in the framework of general quasiparticle approach.

Even the definition of the quasiparticle in SEC systems is not simple and trivial. In the literature one can meet such notions as a spinless fermion, a slave boson and a slave fermion, a holon and a spinon, all of them are reflections of the complicated nature of the quasiparticle in the multielectron system with SEC. In this book we use a more general notion of a

Fermi-type quasiparticle which is a synthesis of the Hubbard ideas and the Landau–Fermi liquid ideas. This notion appears naturally in the atomic representation of the multielectron systems using Hubbard X-operators. One of the main advantages of such an approach is exact counting of the strongest Coulomb matrix elements. Another equally important advantage of the X-operator representation is the exact account of the complicated structure of the Hilbert space at all steps of calculations. The complicated algebra of X-operators automatically fulfills the constraint condition that forbids some sectors of the Hilbert space due to SEC. At the same time, this complicated algebra of X-operators is rather awkward in comparison to the Fermi and Bose-type operators, it requires some experience to work in the X-operator representation. The authors of this book have such experience and it is given in detail here, especially the diagram technique for X-operators.

The book can be divided into two parts. The first part consists of Chapters 1–5. Elementary introduction to the Hubbard model in Chapter 1 is supplemented by the definition of the Hubbard X-operators in a more general way then in the original Hubbard's papers. This general definition is used in the Chapter 2 to rewrite the most popular multielectron models in the X-operators representation and to show that all these models are equivalent to the multiorbital generalized Hubbard model. In Chapter 3, ideas of Landau and Hubbard are combined to define the Fermi- and Bose-type quasiparticles in SEC systems, the X-operators provide a natural mathematical tool for this definition. The generalized tight-binding method for the quasiparticle band structure is developed taking into account the SEC. We show that this method is a perturbative realization of the exact Lehmann representation for the electron Green function. In this method exact diagonalization of the intracell multielectron Hamiltonian is used and Chapter 4 describes the unitary transformations in the X-operator representation. Chapter 5 gives an introduction to the diagram technique for the X-operators. This technique is more complicated and not so widely used as the Fermi- or Bose-operators technique. Nevertheless we describe here this technique in detail so that it is possible for the readers to use it after reading this chapter.

The second part of the book is given by Chapters 6–8. Here we demonstrate how the general method developed in the first part works for several examples: rare-earth or actinide metallic ferromagnets with strong single-ion anisotropy (Chapter 6), non-Fermi liquid effects in the de Haas-van

Alphen oscillations in magnetically ordered SEC systems (Chapter 7) and the electronic structure of copper oxides (Chapter 8).

The authors would like to point out the salient features that distinguish this work from other books on this subject (Fulde 1991; Gebhard 1997; Izyumov et al 1994; Fazekas 1999): It is the novel application of the X-operators representation in a wide variety of different models.

This book can be useful for a wide circle of researchers and as a textbook for an advanced course for graduate and postgraduate students working in the field of SEC systems applied to the problems of magnetism, superconductivity, electronic phase transitions and electron spectroscopy of solids. Parts 1 and 2 of this book have been used in the advanced physics course for the graduate students in Krasnoyarsk State University and Omsk State University. It is our pleasure to thank V. A. Gavrichkov, E. V. Kuzmin and I. S. Sandalov for helpful discussions, T. M. Ovchinnikova, and N. V. Lishneva for technical assistance in the preparation of the book. One of us (SGO) thanks D. M. Edwards and the Sir Issac Newton Institute for Mathematical Sciences for the hospitality during the visit to Cambridge that was very fruitful for the writing of this book. Some parts of the book have been written under the financial support of the Russian Federal Program "Integration of fundamental research and high school education" grants A0019, B0017, Russian Foundation for Basic Researches grants 99-02-17405, 00-02-97705, 02-02-16110 03-02-16124, INTAS grant 01-0654, and "Quantum macrophysics" program of Russian Academy of Sciences.

S. G. Ovchinnikov
V. V. Val'kov

Chapter 1

Hubbard Model as a Simplest Model of Strong Electron Correlations

The Hubbard model can be considered as a minimal model taking into account inter-electron interactions as well as their kinetic energy. The model is obviously oversimplified to give a quantitative description of specific substances. Still, the Hubbard model is one of the important models of modern solid state theory. In spite of the simplicity of the Hubbard model, the model Hamiltonian is able to describe a number of such nontrivial phenomena as metal–insulator transition, ferromagnetism and antiferromagnetism, superconductivity, and Luttinger quantum liquid.

1.1 Hamiltonian of the Hubbard Model and its Symmetry

The Hubbard model is a particular limit of a general model of interacting electrons with the band structure given in the tight-binding approach (Fig. 1.1).

As compared to the general tight-binding model of electrons with Coulomb interaction the Hubbard model has two important simplifications. Firstly, only one electron orbital at each site is considered and all other orbitals are assumed to be unimportant in the low energy physics. Secondly, account is made of the intra-atomic Coulomb matrix element U only. The Hamiltonian of the Hubbard model is given by

$$\hat{H} - \mu \hat{N}_e = \sum_{f\sigma} \left[(\varepsilon - \mu) n_{f\sigma} + \frac{1}{2} U n_{f\sigma} n_{f,-\sigma} \right] + \sum_{fg\sigma}(t_{fg} a^+_{f\sigma} a_{g\sigma} + \text{h.c.}),$$

(1.1)

where $n_f = a^+_{f\sigma} a_{f\sigma}$, $a_{f\sigma}$ is the annihilation operator of the electron at site f with spin projection $\sigma = \pm 1/2$, ε is the single electron energy in

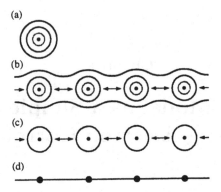

Fig. 1.1 Ideology of the Hubbard model: (a) An atom with several electron orbitals; (b) In a periodic lattice of atoms the external electrons become itinerant while the internal valence electrons are mainly localized with the small probability of interatomic tunnelling; (c) Only one valence orbital — most important for the low energy physics — is considered, all others are neglected; (d) Instead of orbitals with some space distribution of the electron, the lattice model appears with the electron determined at the lattice site with intersite hopping probability.

the crystal field, μ is the chemical potential, U is the Coulomb repulsion parameter, and t_{fg} is the hopping parameter between sites f and g.

To consider the symmetry of the Hubbard model we define a spin operator at site f:

$$\hat{S}_f^{(\alpha)} = \sum_{\sigma\sigma'} a_{f\sigma}^+ (\tau^{(\alpha)})_{\sigma\sigma'} a_{f\sigma'}, \qquad (1.2)$$

with $\alpha = x, y, z$ and $\tau^{(\alpha)}$ being the Pauli matrices. The total spin of the system is equal to

$$\hat{S}_{\text{tot}}^{(\alpha)} = \sum_f \hat{S}_f^{(\alpha)}.$$

The operators $\hat{S}_{\text{tot}}^{(\alpha)}$ commute with both the kinetic and Coulomb parts of Hamiltonian (1.1), that means they are invariance of the Hubbard model Hamiltonian under arbitrary global rotations in the spin subspace. The quantum numbers are the z-projection of S_{tot}^z and the value of total spin $S_{\text{tot}}(S_{\text{tot}} + 1)$ with the total spin operator

$$\hat{S}_{\text{tot}}^2 = \hat{S}_{\text{tot}}^{(x)^2} + \hat{S}_{\text{tot}}^{(y)^2} + \hat{S}_{\text{tot}}^{(z)^2}.$$

For a given number of electrons N_e and sites N, one can found a maximal possible value of the total spin

$$S_{\max} = \begin{cases} N_e/2 & 0 \leq N_e \leq N \\ N - N_e/2 & N \leq N_e \leq 2N \end{cases}.$$

Possible values of S_{tot} are given by

$$S_{\text{tot}} = 0, 1, \ldots, S_{\max} \quad (S_{\text{tot}} = 1/2, 3/2, \ldots, S_{\max})$$

for even (odd) number of electrons N_e.

1.2 Time Inversion

A grand canonical partition function for the Hubbard model is trivially invariant under the phase transformations of the electron creation and annihilation operators. Simultaneously, the single electron wavefunctions in the Wannier representation transform like

$$\varphi_f(\boldsymbol{r}) \to \varphi_f(\boldsymbol{r}) \exp(i\alpha_f).$$

The choice $\alpha_f - \alpha_g = \pi$ for nearest neighbors f, g, with the nonzero hopping parameter t_{fg}, results in the equation

$$Z_\mu(t) = Sp\, e^{-\beta(\hat{H} - \mu \hat{N}_e)} = Z_\mu(-t), \tag{1.3}$$

meaning that it is an invariance of the partition function under the sign inversion of the hopping parameter.

1.3 Electron Hole Symmetry

The hole representation is given by the transformation

$$a_{f\sigma}^+ \to a_{f,-\sigma},$$

resulting in the following transition of the Hamiltonian

$$H(t) - (\mu - \varepsilon)N_e \to H(-t) - [U - (\mu - \varepsilon)]N_e + (2\varepsilon - 2\mu + U)N.$$

The partition functions in the electron and the hole representations are connected via the relation:

$$Z_\mu(t) = \exp[-\beta(U - 2\mu + 2\varepsilon)N] Z_{U-\mu}(t).$$

Taking the derivative of μ in both parts of this equation, one gets

$$\frac{1}{\beta} \cdot \frac{\partial}{\partial (U-\mu)} \ln Z_{U-\mu}(t) = 2N - \langle N_e \rangle.$$

This equation means that in the half-filled $\langle N_e \rangle = N$ case, the chemical potential is equal to (Castellani et al 1979):

$$\mu = \varepsilon + U/2, \qquad (1.4)$$

and this equation is exact for all model parameters and temperatures.

1.4 Pseudospin Symmetry for Half-Filled Case

In the case $\langle N_e \rangle = N$ and for the alternant lattice,[1] the Hubbard model has additional pseudospin isotropy (Yang 1989). The pseudospin vector is defined similar to Eq. (1.2):

$$\boldsymbol{R} = \frac{1}{2} \Psi_\alpha^+(r) \boldsymbol{\tau}_{\alpha\beta} \Psi_\beta(r), \qquad (1.5)$$

where $\boldsymbol{\tau} = (\tau^{(x)}, \tau^{(y)}, \tau^{(z)})$ is the vector constructed with the Pauli matrices and the two component spinor $\Psi_\alpha(r)$ is given by

$$\Psi(r) = \begin{pmatrix} a_\uparrow(r) \\ \exp(iQ \cdot r) a_\downarrow^+(r) \end{pmatrix}. \qquad (1.6)$$

Here, vector $Q = (\pi, \pi, \ldots)$ is a d-dimensional vector with all components equal to π. The explicit form of the pseudospin operator is as follows:

$$R_x(r) = \frac{1}{2} \exp(iQr)(a_\uparrow^+(r) a_\downarrow^+(r) + a_\downarrow(r) a_\uparrow(r)),$$

$$R_y(r) = \frac{1}{2} \exp(iQr)(a_\uparrow^+(r) a_\downarrow^+(r) - a_\downarrow(r) a_\uparrow(r)),$$

$$R_z(r) = \frac{1}{2}(n_\uparrow(r) + n_\downarrow(r) - 1).$$

These operators form an $SU(2)$ algebra and follow the usual spin commutation rules. The total pseudospin operators of the system are equal to

$$R_{\text{tot}}^\pm = \sum_r (R_x(r) \pm iR_y(r)), \qquad R_{\text{tot}}^z = \sum_r R_z(r), \qquad (1.7)$$

---

[1] Lattice that can be separated in two sublattices A and B with each site A having all nearest neighbors from B and vice versa.

with

$$R^2_{\text{tot}} = \frac{1}{2}(R^+_{\text{tot}} R^-_{\text{tot}} + R^-_{\text{tot}} R^+_{\text{tot}}) + (R^z_{\text{tot}})^2.$$

Commutating these with the Hamiltonian, we have

$$[\hat{H}, R^z_{\text{tot}}]_- = [\hat{H}, R^2_{\text{tot}}]_- = 0, \qquad [\hat{H}, R^\pm_{\text{tot}}] = \pm(2\varepsilon + U - 2\mu)R^\pm_{\text{tot}}, \quad (1.8)$$

which show the conservation of R^2_{tot} and R^z_{tot}. In the half-filled case with $\mu = \varepsilon + U/2$, the additional conservation law for $R^\pm_{\text{tot}}$ appears. A combination of the spin and the pseudospin rotation symmetry results in the symmetry group $SO(4)$ (Yang 1989; Zhang 1990).

1.5 The Band Limit

With $U = 0$, the Hamiltonian (1.1) can be diagonalized with the help of the Fourier transformation

$$a_{k\sigma} = \frac{1}{\sqrt{N}} \sum_f a_{f\sigma} \exp(-i\boldsymbol{k} \cdot \boldsymbol{R}_f),$$

from which we find

$$H_0 = \sum_{k\sigma}(\varepsilon(k) - \mu)a^+_{k\sigma} a_{k\sigma}, \qquad (1.9)$$

$$\varepsilon(k) = \varepsilon + t(k), \qquad t(k) = \sum_g t_{fg} \exp(i\boldsymbol{k} \cdot (\boldsymbol{R}_f - \boldsymbol{R}_g)).$$

Hamiltonian H_0 describes a system of free electrons with bandwidth $2W$. For the case where there is only nearest neighbor hopping, the value of half bandwidth is equal to $W = |t|z$ with the number of neighboring sites z. At small concentration of electrons ($n_e \ll 1$) or holes ($n_h = 2 - n_e \ll 1$), the effective mass approximation to the electron spectrum is justified, resulting in a spherical Fermi surface

$$p_x^2 + p_y^2 + p_z^2 = 2m * \varepsilon_F.$$

In general case, with arbitrary dependence of $\varepsilon(p)$ on $\boldsymbol{P}$, the form of the Fermi surface may be rather complicate. For example, in two dimensional square lattice in the tight binding method, with account only for nearest neighbor hopping t_{ij}, the Fermi surface at $n_e = 1$ is a square given by the equation

$$\cos(k_x a) + \cos(k_y a) = 0. \tag{1.10}$$

In the weak-coupling limit $U \ll W$, a rather good approximation is given by the Hartree–Fock solution with the decoupling of the Coulomb part of Eq. (1.1):

$$U n_{f\uparrow} n_{f\downarrow} \to U n_{f\uparrow} \langle n_{f\downarrow} \rangle + U \langle n_{f\uparrow} \rangle n_{f\downarrow}.$$

The Hartree–Fock band structure depends on the electron occupation numbers. In the case where the electron is uniformly distributed, it is given by

$$\varepsilon_{\mathrm{HF}}(k, \sigma) = \varepsilon(k) + U \langle n_{-\sigma} \rangle. \tag{1.11}$$

Due to the electron occupation numbers, the band structure depends on the magnetic field, temperature and electron concentration $n_e = \langle N_e \rangle / N$. In principle, there are several possible magnetic solutions: ferromagnetic, ferrimagnetic and antiferromagnetic. The antiferromagnetic phase for the half-filled case describes an insulator (static spin density wave (SDW) state) with dispersion of electrons at the bottom of the empty conductivity band $E^+(k)$ and at the top of the filled valence band $E^-(k)$, given by

$$E^\pm(k) = \varepsilon \pm (t(k)^2 + U^2 \langle S^z \rangle^2)^{1/2}, \tag{1.12}$$

where $\langle S_g^z \rangle$ is the sublattice magnetization. The metal-insulator transition in the SDW state in the band limit has been presented in details in several reviews (Mott 1974; Izyumov et al 1994) and we will not consider it here.

1.6 The Atomic Limit, Hubbard 1 Solution

In the absence of hopping, $t_{fg} = 0$, the model describes a system of individual sites, each can be in one of the following eigenstates:

$$|0\rangle \equiv |\mathrm{vacuum}\rangle; \quad |\sigma\rangle = a_{f\sigma}^+ |0\rangle, \quad \sigma = \pm 1/2; \quad |2\rangle = a_{f,+}^+ a_{f,-}^+ |0\rangle \tag{1.13}$$

with eigenvalues $\varepsilon_0 = 0$, $\varepsilon_\sigma = \varepsilon - \mu \equiv \varepsilon_1$, $\varepsilon_2 = 2(\varepsilon - \mu) + U$. The number of electrons per site is given by

$$n = \frac{1}{N} \sum_f \langle \hat{N}_f \rangle = \frac{1}{N} \sum_{f\sigma} \langle a_{f\sigma}^+ a_{f\sigma} \rangle = N_+ + N_- + 2N_2, \tag{1.14}$$

where N_+, N_-, N_2 are the filling numbers of the eigenstates $|1/2\rangle$, $|-1/2\rangle$, $|2\rangle$. The energy of the system is equal to

$$E = \sum_{f\sigma} \varepsilon_1 N_{f,\sigma} + \sum_f \varepsilon_2 N_{f,2}. \tag{1.15}$$

The filling numbers are determined self-consistently via the chemical potential equation. The ground state of the system is highly degenerate. A weak intersite hopping eliminates this degeneration.

To obtain the quasiparticle spectrum one should calculate the single electron Green's function

$$G_\sigma(\boldsymbol{k}, E) = \langle\langle a_{\boldsymbol{k}\sigma}|a^+_{\boldsymbol{k}\sigma}\rangle\rangle_E$$
$$= \sum_{(\boldsymbol{R}_f - \boldsymbol{R}_{f'})} \exp(-i\boldsymbol{k} \cdot (\boldsymbol{R}_f - \boldsymbol{R}_{f'})) \langle\langle a_{f\sigma}|a^+_{f'\sigma}\rangle\rangle_E. \tag{1.16}$$

An exact equation of motion for the Green's function is given by

$$(E - \varepsilon + \mu)\langle\langle a_{f,\sigma}|a^+_{f'}\rangle\rangle = \delta_{ff'} + \sum_h \langle\langle a_{f+h,\sigma}|a^+_{f'}\rangle\rangle t(h) + U F_\sigma(f - f', E),$$
$$F_\sigma(f - f', E) = \langle\langle a_{f,\sigma} n_{f,-\sigma}|a^+_{f'}\rangle\rangle, \tag{1.17}$$

and involves a higher-order Green's function F. The exact equation of motion for the F function has form

$$(E - \varepsilon + \mu)\langle\langle a_{f,\sigma} n_{f,-\sigma}|a^+_{f',\sigma}\rangle\rangle$$
$$= \delta_{ff'} \langle n_{f,-\sigma}\rangle + U\langle\langle a_{f,\sigma} n_{f,-\sigma}|a^+_{f',\sigma}\rangle\rangle + \sum_h t(h)(\langle\langle a_{f+h,\sigma} n_{f,-\sigma}|a^+_{f',\sigma}\rangle\rangle$$
$$+ \langle\langle a^+_{f,-\sigma} a_{f+h,-\sigma} a_{f,\sigma}|a^+_{f',\sigma}\rangle\rangle + \langle\langle a^+_{f-h,-\sigma} a_{f,\sigma} a_{f,-\sigma}|a^+_{f',\sigma}\rangle\rangle). \tag{1.18}$$

A simplest decoupling in the last line of Eq. (1.18) is known as the Hubbard 1 decoupling (Hubbard 1963):

$$\langle\langle a_{f+h,\sigma} n_{f,-\sigma}|a^+_{f',\sigma}\rangle\rangle \to \langle n_{f,-\sigma}\rangle \langle\langle a_{f+h,\sigma}|a^+_{f',\sigma}\rangle\rangle.$$

In the Hubbard 1 approximation, the system of equations for functions G and F is closed and after the Fourier transformation it may be written as

$$(E - \varepsilon(k) + \mu)G_\sigma(k, E) = \mathbb{1} + U F_\sigma(k, E),$$
$$(E - \varepsilon - U + \mu)F_\sigma(k, E) = \langle n_{-\sigma}\rangle + \langle n_{-\sigma}\rangle t(k) G_\sigma(k, E). \tag{1.19}$$

Here, the space uniform solution $\langle n_{f,-\sigma}\rangle = \langle n_{-\sigma}\rangle$ is assumed. The single electron Green's function (1.16) is given by

$$G_\sigma(k, E) = \frac{E - \varepsilon - U(1 - \langle n_{-\sigma}\rangle) + \mu}{(E - E_1(k))(E - E_2(k))}, \qquad (1.20)$$

$$E_{1,2}(k) = \varepsilon + \frac{U + t(k)}{2} \pm \left[\left(\frac{U - t(k)}{2}\right)^2 + Ut(k)\langle n_{-\sigma}\rangle\right]^{1/2} - \mu. \qquad (1.21)$$

In the paramagnetic half-filled case $\langle n_{-\sigma}\rangle = 1/2$, $\mu = \varepsilon + U/2$, the quasiparticle spectrum has form

$$E_{1,2}(k) = \frac{1}{2}\left(t(k) \pm (U^2 + t^2(k))^{1/2}\right). \qquad (1.22)$$

The subbands $E_1(k)$ and $E_2(k)$ are called the Hubbard subbands. The Hubbard 1 solution has been subjected to criticism in literature because it contains a gap between $E_1(k)$ and $E_2(k)$ subbands for all values of U while the metal–insulator transition and the disappearance of the gap are intuitively expected at $U \sim W$. Nevertheless in the atomic limit $U \gg W$ the Hubbard 1 solution is a good starting approximation. Further improvement of the Hubbard 1 solution as it is apparent from the last line in Eq. (1.18) involves the correlation functions of the charge density, the spin density and the pair density. Several approximations have been developed to obtain the metal–insulator transition: Hubbard 3 (Hubbard 1964), Gutzwiller approximation (Gutzwiller 1963; 1964; 1965), coherent potential approximation (Velicky et al 1968), single-loop approximation in the X-operator diagram technique (Zaitsev 1976), slave-boson approach (Kotlair and Ruchenstein 1986), and dynamical mean-field approximation in the infinite dimension (Metzner and Vollhardt, 1989).

One important distinction of the Hubbard subbands is the dependence of the number of states in a given subband on the electron occupation numbers, while for the conventional single electron bands the total number of states in each band is equal to 2, taking the spin into account. To prove this peculiarity we introduce the single electron spectral density function

$$A_\sigma(k, E) = -\frac{1}{\pi}\mathrm{Im}\, G_\sigma(k, E + i\delta)$$

which, in Hubbard 1 approximation together with the Green's function (1.20), is given by a sum of lower and upper Hubbard band contributions

$$A_\sigma(k, E) = \frac{E_1(k) - \varepsilon - U(1 - \langle n_{-\sigma} \rangle) + \mu}{E_1(k) - E_2(k)} \delta(E - E_1(k))$$

$$+ \frac{U(1 - \langle n_{-\sigma} \rangle) + \varepsilon - E_2(k) - \mu}{E_1(k) - E_2(k)} \delta(E - E_2(k))$$

$$\equiv A_{1\sigma}(k, E)\delta(E - E_1(k)) + A_{2\sigma}(k, E)\delta(E - E_2(k)). \quad (1.23)$$

It is clear from Eq. (1.23) that $A_{1\sigma} + A_{2\sigma} = 1$, i.e. only the sum from both bands will be equal to 2. For the separate bands, the density of states is equal to

$$N_{1\sigma,2\sigma}(E) = \frac{1}{N} \sum_k A_{1\sigma,2\sigma}(k, E).$$

Integration over energy results in (Fulde 1991):

$$N_{1\sigma} = \int N_{1\sigma}(E) dE = \langle n_{-\sigma} \rangle, \qquad N_{2\sigma} = \int N_{2\sigma}(E) dE = 1 - \langle n_{-\sigma} \rangle.$$
(1.24)

For the half-filled case $N_{1\sigma} = N_{2\sigma} = 1/2$, the total number of states in each Hubbard band, taking spin into account, is 1. This fact is often used to prove that an electron in the limit $U \to \infty$ becomes a spinless particle. It is not true because there is a 2-fold spin degeneracy in the paramagnetic phase and the Zeeman splitting takes place in the external magnetic field. The situation with the quasiparticle statistics is more complicate and will be discussed below in the Chapter 3.

1.7 The Hubbard Model in One and Infinite Dimension Cases

For the one-dimensional case $d = 1$, there is an exact solution by Lieb and Wu (1968). The ground state energy is given for the half-filled case $n_e = 1$ by

$$\frac{E_0}{N} = -4|t| \int_0^\infty \frac{dx\, J_0(x) J_1(x)}{x[1 + \exp(xU/|t|)]}, \quad (1.25)$$

where $J_n(x)$ is the Bessel function. The ground state energy in the band limit $U \ll t$ is

$$\frac{E_0}{N} = -4\frac{|t|}{\pi} + \frac{U}{4} - 0.017\frac{U^2}{|t|}; \qquad (1.26)$$

In the atomic limit $U \gg t$, it is

$$\frac{E_0}{N} = -\frac{4|t|^2}{U}\ln 2. \qquad (1.27)$$

The last expression is exactly the energy of $d = 1$ Heisenberg $S = 1/2$ antiferromagnetic chain with the intersite exchange interaction $J = 4t^2/U$.

The ground state for $n_e = 1$ is an insulator for all U/t values. It is proved by the fact that the chemical potential required to add one electron μ_+ is larger than the potential to remove one electron μ_-. Relation $\mu_+ > \mu_-$ is typical for insulator with the energy gap between the filled and empty bands.

The exact wavefunction of the ground state is given by the Bethe-anzatz and has rather a complicated structure that may be simplified in the limit $U \to \infty$ when all two-electron states at one site can be excluded. It results in the following wavefunction (Ogato and Shiba 1990):

$$\Psi(x_1, \ldots, x_N) = \det|\exp(ik_j x_j)|\Phi(y_1, \ldots, y_M), \qquad (1.28)$$

where $x_1, \ldots, x_M$ are the spin-up electron sites, $x_{M+1}, \ldots, x_N$ are the spin-down electron sites. In the right side of Eq. (1.28), the Sleter determinant describes the noninteracting spinless fermions with moment $k_1, \ldots, k_N$ and the function Φ in Eq. (1.28) is the exact wavefunction of the Heisenberg $d = 1$ chain with $S = 1/2$ where $y_1, \ldots, y_M$ are the "pseudocoordinates" of the spin-down particles.

Factorization of the wavefunction (1.28) into the product of two factors with one corresponding only to the charge degree of freedom and the other corresponding only to the spin variables is a very important result. Sometimes it is claimed to be the general property of SEC systems. It is not clear whether this result is true for lattice with dimension $d > 1$. At least another nontrivial solution of the Hubbard model in infinite dimension $d \to \infty$ gives no indication for the charge-spin separation at the level of one electron. The $d = 3$ systems seems to be appropriate to treat in the $1/d$ expansion. The most unclear situation is for $d = 2$ systems.

The low energy excitations in the $d = 1$ Hubbard model have been studied in detail. Instead of producing the Fermi liquid, the SEC effects result in the Luttinger liquid (Tomonaga 1950; Luttinger 1968; Solyom 1979; Haldane 1981). For the Luttinger liquid the function $n_k = \langle a_k^+ a_k \rangle$ has no jump at the Fermi level,

$$n_k - n_{kF} \sim \text{sign}(k - k_F)|k - k_F|^\alpha$$

where α is a nonuniversal constant and the value of α is determined by the interaction constant. In the limit $U \to \infty, \alpha = 1/8$ (Parolla and Sorella 1990). Instead of having the delta-function (as in the Fermi liquid), the spectral density in the Luttinger liquid near the Fermi level has the form $A(\omega) \sim |\omega - \varepsilon_F|^\alpha$. Zero spectral density at ε_F in the one-dimensional Hubbard model was the main reason to abandon the description of electron as a particle with charge e and spin $1/2$ in SEC systems (Anderson 1990). Meanwhile the QMC calculations (Preuss et al 1994) have revealed the quasiparticle peak in $A(k, \omega)$ typical for the Fermi liquid. In the limit $U = \infty$ the spectral density $A(k, \omega)$ has been calculated (Penc et al 1995) using the exact wavefunction (1.28). The multielectron effects decrease the van Hove singular contributions to $A(k, \omega)$ at the band edge and a new singularity $A(\omega) \sim |\omega - \varepsilon_F|^{-3/8}$ (for electron concentration $n_e < 1$) has been found (Fig. 1.2). The non-quasiparticle tails in the spectral density above the bandwidth are also the result of SEC.

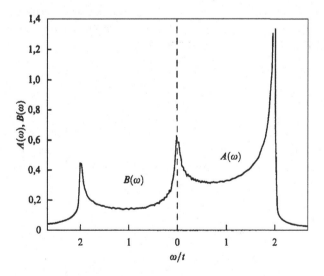

Fig. 1.2 The density of states of the lower Hubbard band for $n_e = 1/2$ and $U = \infty$ for the one-dimensional Hubbard model (Reprinted with permission from Penc et al, Phys. Rev. Lett. **75**, 894 (1995). © (1995) American Physical Society).

Main contributions to the $A(k,\omega)$ (more than 99.99%) are given by the sum of the single particle excitations and the three particle excitations involving the electron–hole pairs. The applicability of the Luttinger liquid ideas has also been analyzed by Penc et al (1995) and they have found $\alpha = 1/8$ in the region $|\omega - \varepsilon_F| < t^2/U$; outside this region $\alpha = -3/8$. When $U \to \infty$, the region of the Luttinger liquid tends to zero; so at large but finite U the spectral density $A(\omega)$ has its peak at $\omega \sim t^2/U$ because of the singularity with $\alpha = -3/8$. This result explains the peak in $A(\omega)$ obtained previously by QMC (Preuss et al 1994).

In the other limit $d \to \infty$ the non-trivial solutions for the Hubbard model have been found by Metzner and Vollhardt (1989). In this limit the self-energy becomes local (Muller-Hartmann 1989):

$$\sum\nolimits_{fg}(\omega) = \sum(\omega)\delta_{fg}. \tag{1.29}$$

This makes it possible to construct a nontrivial mean field theory (which is exact in the $d \to \infty$ limit) that covers the main results expected in $d = 3$: the metal–insulator transition, the antiferromagnetic order with the local magnetic moments in the half-filled case, the Fermi liquid behavior in the metal phase with heavy fermions with the mass increasing near the border of localization. There are several reviews devoted to the $d = \infty$ limit (Izyumov 1995; Georges et al 1996; Gebhard 1997). Some exact results for the Hubbard model are presented by Tasaki (1998).

1.8 The t–J Model

This model have been derived from the Hubbard model in the limit $t/U \ll 1$. Using canonical transformation which excludes the local two-electron states we get the following Hamiltonian (Bulayevsky et al 1968; Hirsch 1985):

$$\begin{aligned}H_{t-J^*} = &-t \sum_{\langle i,j\rangle \sigma}(c^+_{i\sigma}c_{j\sigma} + c^+_{j\sigma}c_{i\sigma}) \\ &+ \frac{t^2}{U}\sum_{j,\delta,\delta'}(c^+_{j+\delta\uparrow}c^+_{j\downarrow}c_{j\downarrow}c_{j+\delta'\uparrow} + c^+_{j\uparrow}c^+_{j+\delta\downarrow}c_{j+\delta'\downarrow}c_{j\uparrow} \\ &+ c^+_{j+\delta\uparrow}c^+_{j\downarrow}c_{j+\delta'\downarrow}c_{j\uparrow} + c^+_{j\uparrow}c^+_{j+\delta\downarrow}c_{j\downarrow}c_{j+\delta'\uparrow}),\end{aligned} \tag{1.30}$$

where $j+\delta'$ are the nearest neighbors to the site j. Neglecting all 3-site terms and taking $\delta' = \delta$, we have the t–J model Hamiltonian with $J = 4t^2/U$

$$H_{t-J} = -t \sum_{\langle i,j\rangle \sigma} (c^+_{i\sigma} c_{j\sigma} + \text{h.c.}) + \sum_{\langle i,j\rangle} J(\boldsymbol{S}_i \cdot \boldsymbol{S}_j - n_i n_j/4). \qquad (1.31)$$

Here the Fermi operators are defined in a restricted Hilbert space without double electron occupations at one site,

$$c_{i\sigma} \to c_{i\sigma}(1 - n_{i,-\sigma}).$$

The 3-site term in Eq. (1.30) makes its contribution only in the case of nonzero number of electrons (or holes). It may be rewritten as follows

$$H^{(3)}_J = -\frac{t^2}{U} \sum_{j,\sigma} \sum_{\delta \neq \delta'} \left(c^+_{j+\delta,\sigma} n_{j,-\sigma} c_{j+\delta',\sigma} - c^+_{j+\delta,\sigma} c^+_{j,-\sigma} c_{j,\sigma} c_{j+\delta',-\sigma} \right), \qquad (1.32)$$

where the first term corresponds to the intrasublattice hopping that does not affect the antiferromagnetic order and the second term describes the hopping between the next nearest neighbors accompanied by the spin fluctuation. Both these hoppings bring forth the so-called t–t'–t''–J model (Belinicher et al 1996).

To estimate the role of 3-site terms in Eq. (1.30) and the accuracy of the t–J model the energy of ground state has been computed by the exact diagonalization method for $d = 1$ chain with 10 atoms and $d = 2$ lattice cluster with 10 atoms in the framework of a few models: the Hubbard model; the t–J model without carriers; the t–J, t–J^* and the Hubbard model with one carrier (von Szczepanski et al 1990). It turns out that at $U/t = 20$ all models give the same value of energy with less than 1% differences. With U/t decreasing, for example at $U/t = 8$, the difference in the $d = 2$ cluster is more than 20% which is a clear indication to take into account higher order t/U terms in the effective Hamiltonian. Strictly speaking the t–J model is deduced from the Hubbard model at $J \ll t$. Nevertheless due to the nontrivial properties of the t–J model, one often considers arbitrary values of J/t. Some justification for the more broader consideration of the t/J model has been done by Eder et al (1996) who showed that the additional Coulomb matrix elements V_{ij} can result in the instability with respect to the charge density wave state. A system close to this instability may be a realization of the t–J model with $J \sim t$.

The quasiparticles in the t–J model have been intensively discussed in literature, see e.g. reviews by Fulde (1991), Dagotto (1994), Izyumov et al

(1994), Kampf (1994), Brenig (1995), and Ovchinnikov (1997). In most approaches the quasiparticle is a spin polaron: it is an electron dressed by the spin fluctuations. The distinction between the free electron and the spin polaron has been demonstrated by Eder and Ohta (1994) where the spectral density $A(k,\omega)$ for the t–J model was calculated by exact diagonalization method. The annihilation operator of polaron can be written in the following form

$$c_{k,\uparrow} = \frac{1}{\sqrt{N}} \sum_j \sum_{\lambda=0}^{3} e^{i\mathbf{k}\cdot\mathbf{R}_j} \alpha_\lambda(k) A_{j,\lambda}, \qquad (1.33)$$

$$A_{j,0} = c_{j,\uparrow}, \quad A_{j,1} = \sum_{h \in N(j)} S_j^- c_{h,\downarrow}, \quad A_{j,2} = \sum_{h \in N(j)} \sum_{l \in N(h)} S_j^- S_h^+ c_{l,\uparrow},$$

$$A_{j,3} = \sum_{h \in N(j)} \sum_{l \in N(h)} \sum_{m \in N(l)} S_j^- S_h^+ S_l^- c_{m,\downarrow}.$$

Here, $N(j)$ denotes a set of nearest neighbors to R_j, and $a_\lambda(k)$ are the variational parameters. For the free electron the spectral density has a wide spectrum for almost all k (Fig. 1.3) of non-coherent origin; this means that the free electron state is not the proper state of the system. Conversely the narrow peaks in $A(k,\omega)$ for the spin polaron mean that it is a well-defined quasiparticle for the t–J model.

1.9 The Hubbard Model in X-Operator Representation

In the atomic limit it is convenient to use the X-operators (Hubbard 1965). It turns out that the X-operator representation is also useful for a broader range of different models (see the next chapter), so we will define X-operators in a more general form then it was done by Hubbard. Let us consider any local Hamiltonian H_f, which may be that of a unit cell centred at site f with several atoms per cell. Let $\{|p\rangle\} = \{|1\rangle, |2\rangle, \ldots, |n\rangle\}$ be a complete set of orthogonal and normalized eigenstates of H_f. This set forms a local basis at the cell f of the Hilbert subspace H_n and all the physical operators corresponding to the given cell are the operators in H_n. The total number of linear independent operators in H_n is equal to n^2. We take these operators in the form of Hubbard X-operators

$$X_f^{pq} = |p\rangle\langle q|, \qquad (1.34)$$

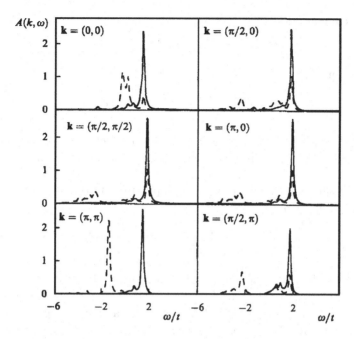

Fig. 1.3 The spectral functions of a free electron (dashed line) and a spin polaron (solid line) for the $t-J$ model with the electron concentration $n_e = 1$ obtained by the exact diagonalization of 4 × 4 cluster. (Reprinted with permission from Eder and Ohta, *Phys. Rev. B* **50**, 10043 (1994). © (1994) American Physical Society.)

where indices p, q denote the corresponding eigenstates of $H_f (1 \leq p \leq n,$ $1 \leq q \leq n)$. By action of X_f^{pq} on the state $|l\rangle$, we have

$$X^{pq}|l\rangle = |p\rangle\langle q|l\rangle = \delta_{ql}|p\rangle. \quad (1.35)$$

From Eq. (1.35), it is obvious that X^{pq} is a transition operator with the initial state $|q\rangle$ and the final state $|p\rangle$. The diagonal operators X_f^{mm} are the projection operators

$$(X^{nn})^2 = X^{nn}X^{nn} = X^{nn}, \quad (1.36)$$

while the nondiagonal operators have no projection properties as $(X^{pq})^2 = 0$. The X-operators algebra is given by the following rules:

(a) the multiplication rule

$$X_f^{p_1 q_1} \cdot X_f^{p_2 q_2} = |p_1\rangle\langle q_1||p_2\rangle\langle q_2| = \delta_{q_1 p_2} X_f^{p_1 q_2},$$

(b) the commutation rule

$$\left[X_f^{p_1q_1} \cdot X_g^{p_2q_2}\right]_\pm = \delta_{fg}\left(\delta_{q_1p_2} X_f^{p_1q_2} \pm \delta_{p_1q_2} X_f^{p_2q_1}\right),$$

(c) the sum rule

$$\sum_p X_f^{pp} = \sum_p |p\rangle\langle p| = 1.$$

All of these rules are evidently the results of the definition of X-operators and the condition that $\{|p\rangle\}$ be a complete set of orthogonal and normalized vectors.

Any local operator $\hat{Q}_f$ in H_n can be rewritten in the X-operators representation using the identity

$$\hat{Q} = \hat{1} \cdot \hat{Q} \cdot \hat{1} = \sum_{pq} |p\rangle\langle p|\hat{Q}|q\rangle\langle q| = \sum_{pq} \langle p|\hat{Q}|q\rangle X^{pq}, \quad (1.37)$$

where the matrix element $\langle p|\hat{Q}|q\rangle$ gives the probability of the system under the action of operator $\hat{Q}$, going from the initial state $|q\rangle$ to the final state $|p\rangle$.

It is not convenient to describe each X-operator by the pair of indices (p, q). To simplify the notations we introduce an ordered set of n^2 elements (p, q), for example as follows

$$(11), (12), \ldots (1n), (21), (22), \ldots, (2n), \ldots, (n1), (n2), \ldots, (nn). \quad (1.38)$$

We make a one-to-one correspondence of this set to the other set of n^2 elements

$$\lambda_1, \lambda_2, \ldots, \lambda_{n^2}. \quad (1.39)$$

Using these notations, we rewrite Eq. (1.37) in the form

$$Q = \sum_\lambda \gamma(\lambda) X^\lambda, \quad \gamma(\lambda) = \langle p|\hat{Q}|q\rangle. \quad (1.40)$$

The coefficients $\gamma(\lambda)$ are the parameters of the X-representation for the Q operator. The single index form (1.40) is more convenient compare to Eq. (1.37) for two reasons. Firstly, it gives a more simple formulation of the Wick theorem for the X-operators which will be considered below in Chapter 5. Secondly, the interaction matrix $V_{\lambda_1\lambda_2}$ turns out to be a product of two factors $f(\lambda_1) \cdot f(\lambda_2)$ which is very important in solving the system of equations for the Green's functions.

All X-operators can be divided into two classes: the quasi-Bose and the quasi-Fermi operators depending on the difference of the fermion numbers

n_p and n_q in the final $|p\rangle$ and the initial $|q\rangle$ states. If $n_p - n_q$ is even, then the X^{pq} operator is of the Bose type, and when $n_p - n_q$ is odd the X^{pq} operator is of the Fermi type. It is obvious that diagonal operators X^{pp} are of the Bose type, we introduce special notation for the diagonal operators

$$h_{fp} = X_f^{pp}, \quad p = 1, 2, \ldots, n.$$

These operators may be considered components of an n-dimensional vector $\mathbf{h}_f$:

$$\mathbf{h}_f \equiv (h_{f1}, h_{f2}, \ldots, h_{fn}).$$

We also define an n-dimensional vectors $\boldsymbol{\alpha}(p,q)$ for nondiagonal operators $(p \neq q)$ with components

$$\alpha_i(p,q) = \delta_{ip} - \delta_{iq}, \quad i, p, q = 1, 2, \ldots, n. \tag{1.41}$$

Such vectors were introduced by Zaitsev (1975) and were called root vectors. We follow here Zaitsev's idea but with some modification in notations so the components of our root vectors (1.41) differ from those of Zaitsev. The difference is caused by the use of the zero spur diagonal operators $X^{pp} - \langle X^{pp}\rangle$ in Zaitsev work. As is shown below, it is more convenient to work with our definition of the diagonal operators (Sandalov and Podmarkov 1980; Izyumov et al 1994).

All these new notations have been introduced to make formulae that are more compact for the commutation rules. Thus, we can write

$$[h_{fn}, X_g^\alpha]_- = \delta_{fg}\alpha_n X_f^\alpha, \tag{1.42}$$

$$[X_f^\alpha, X_g^{-\alpha}]_\pm = \delta_{fg}(\boldsymbol{\alpha}, \hat{A}_\alpha \mathbf{h}_f), \tag{1.43}$$

where a "metric" matrix $\hat{A}_\alpha$ has been introduced with components

$$(\hat{A}_\alpha)_{ij} = \delta_{ij}\{\delta_{ip} + (-1)^{\Delta_{pq}}\delta_{iq}\}. \tag{1.44}$$

Here p and q are given by the root vector $\alpha(p,q)$ and $\Delta_{pq} = n_p - n_q$ is the change of the fermion numbers between the initial and final states. From Eq. (1.44), it is obvious that for the Bose type operators, the commutator is given by

$$[X_f^\alpha, X_g^{-\alpha}]_- = \delta_{fg}(\boldsymbol{\alpha} \cdot \boldsymbol{h}_f) = \delta_{fg}(h_{fp} - h_{fq}), \quad \boldsymbol{\alpha} = \boldsymbol{\alpha}(p,q), \qquad (1.45)$$

and for the Fermi type operators, the anticommutator is

$$[X_f^\alpha, X_g^{-\alpha}]_+ = \delta_{fg}(h_{fp} + h_{fq}), \quad \boldsymbol{\alpha} = \boldsymbol{\alpha}(p,q). \qquad (1.46)$$

For two different Fermi type operators X_f^α and X_f^β (with $\beta \neq \alpha$), the anticommutator is given by

$$[X_f^\alpha, X_g^\beta]_+ = \delta_{fg} N_{\alpha+\beta} X_f^{\alpha+\beta}, \qquad (1.47)$$

where

$$N_{\alpha+\beta} = \begin{cases} 1, & \text{if vector } \boldsymbol{\alpha} + \boldsymbol{\beta} \text{ is one of the root vectors} \\ 0, & \text{otherwise} \end{cases}.$$

To check whether a vector $\alpha + \beta$ is the root vector is very simple: according to Eq. (1.41) it has only 2 non-zero components, one is $+1$ and the other is -1. When at least one of the different operators X_f^α and X_f^β is of the Bose type their commutator is

$$[X_f^\alpha, X_f^\beta]_- = \pm \delta_{fg} N_{\alpha+\beta} X_f^{\alpha+\beta}. \qquad (1.48)$$

To choose the sign in Eq. (1.48) is easy for every given α and β vectors.

Coming back to the Hubbard model where the atomic basis consists of 4 states $|0\rangle, |\sigma\rangle$ ($\sigma = \pm 1/2$) and $|2\rangle$ [Eq. (1.13)], the X-operator X_f^{pq} can be represented by a 4×4 matrix with all elements equal to zero except one, which is equal to 1, on the intersection of p-line and q-column. There are 8 operators of Fermi type

$$X^{0\sigma}, X^{\sigma 0}, X^{\sigma 2}, X^{2\sigma},$$

4 nondiagonal Bose type operators

$$X^{+-}, X^{-+}, X^{20}, X^{02},$$

and 4 diagonal Bose type operators.

$$X^{00}, X^{++}, X^{--}, X^{22}.$$

For the Hubbard model the X-operators can be expressed via the Fermi operators $a_{f\sigma}$:

$$X_f^{0,0} = (1 - n_{f\uparrow})(1 - n_{f\downarrow}), \quad X_f^{\sigma,\sigma} = n_{f,\sigma}(1 - n_{f,-\sigma}), \quad X_f^{2,2} = n_{f\uparrow}n_{f\downarrow},$$

$$X_f^{\sigma,0} = a_{f,\sigma}^+(1 - n_{f,-\sigma}), \quad X_f^{2,\sigma} = 2\sigma a_{f,-\sigma}^+ n_{f,\sigma},$$

$$X_f^{\sigma,-\sigma} = a_{f,\sigma}^+ a_{f,-\sigma}, \quad X_f^{2,0} = 2\sigma a_{f,-\sigma}^+ a_{f,\sigma}^+.$$

Vice versa, the Fermi operators according to Eq. (1.37) may be written in the X-operator representation as

$$a_{f,\sigma}^+ = X_f^{\sigma,0} + 2\sigma X_f^{2,-\sigma}. \tag{1.49}$$

The number operator $\hat{N}$ of electrons is given by

$$\hat{N} = \sum_{f,\sigma} a_{f\sigma}^+ a_{f\sigma} = \sum_f \left(0 \cdot X_f^{0,0} + 1 \cdot (X_f^{+,+} + X_f^{-,-}) + 2 \cdot X_f^{2,2}\right). \tag{1.50}$$

The Hubbard model Hamiltonian in the X-representation has the form

$$H = H_0 + H_{\text{int}},$$

$$H_0 = \sum_f \left\{ (\varepsilon - \mu) \sum_\sigma X_f^{\sigma,\sigma} + (2\varepsilon + U - 2\mu) X_f^{2,2} \right\}$$

$$H_{\text{int}} = \sum_{f,g} t_{f,g}^{p_1 q_1, p_2 q_2} X_f^{p_1 q_1} X_g^{q_2 p_2}$$

$$= \sum_{f,g} t_{f,g} \{ (X_f^{+,0} + X_f^{2,-})(X_g^{+,0} + X_g^{-,2})$$

$$+ (X_f^{-,0} - X_f^{2,+})(X_g^{0,-} - X_g^{+,2}) \}. \tag{1.51}$$

The Coulomb term in the Hamiltonian (1.1) that is a product of 4 Fermi operators becomes linear in the X-operators while the hopping term is a quadratic form in X-operators. The representation of the Fermi operator (1.49) as a sum of two X-operators corresponds to the decoupling of the free electron band onto two Hubbard subbands. To prove this statement we write down the exact equations of motion for operators in the Heisenberg representation

$$i\dot{X}_f^{0,\sigma} = [X_f^{0,\sigma}, H]_- = \varepsilon_1 X_f^{0,\sigma} + \sum_g t_{f,g} \{ (X_f^{0,0} + X_f^{\sigma,\sigma})(X_g^{0,\sigma} + 2\sigma X_g^{-\sigma,2})$$

$$+ X_f^{-\sigma,\sigma}(X_g^{0,-\sigma} - 2\sigma X_g^{\sigma,2})$$

$$+ 2\sigma X_f^{0,2}(X_g^{-\sigma,0} - 2\sigma X_g^{2,\sigma}) \}, \tag{1.52}$$

$$2i\sigma \dot{X}_f^{-\sigma,2} = (\varepsilon_1 + U)2\sigma X_f^{-\sigma,2} + \sum_g t_{f,g}\{(X_f^{-\sigma,-\sigma} + X_f^{2,2})(X_g^{0,\sigma} + 2\sigma X_g^{-\sigma,2})$$
$$- X_f^{-\sigma,\sigma}(X_g^{0,-\sigma} - 2\sigma X_g^{\sigma,2}) - 2\sigma X_f^{0,2}(X_g^{-\sigma,0} - 2\sigma X_g^{2,\sigma})\}. \tag{1.53}$$

For the Green's functions

$$G_{11}(f, f', E) = \langle\langle X_f^{0,\sigma} | X_{f'}^{0,\sigma} \rangle\rangle_E, \quad G_{21}(f, f', E) = \langle\langle 2\sigma X_f^{-\sigma,2} | X_{f'}^{0,\sigma} \rangle\rangle_E \tag{1.54}$$

in the Hubbard 1 approximation, we have the following system of equations

$$\begin{aligned}(E - \varepsilon_1 - F_{0\sigma}t(k))G_{11}(k, E) - F_{0,\sigma}t(k)G_{21}(k, E) &= F_{0,\sigma}, \\ -F_{-\sigma,2}t(k)G_{11}(k, E) + (E - \varepsilon_1 - U - F_{-\sigma,0}t(k))G_{21}(k, E) &= 0.\end{aligned} \tag{1.55}$$

Here $F_{0\sigma} = \langle X^{00} + X^{\sigma\sigma} \rangle, F_{-\sigma,2} = \langle X^{-\sigma,-\sigma} + X^{22} \rangle$ are the filling factors. From the sum rule, we have $F_{0\sigma} + F_{-\sigma,2} = 1$. It is apparent that the quasiparticle spectrum given by the solution of Eq. (1.55) is described by Eq. (1.21). Moreover in the limit $U \gg t$ when the intersubband hopping in Eq. (1.51) is of minor importance we can write the solution of Eq. (1.55) for the lower Hubbard band (LHB) as:

$$G_{11}^{(0)}(k, E) = \frac{F_{0\sigma}}{E - \varepsilon_1 - F_{0\sigma}t(k)}, \tag{1.56}$$

and for the upper Hubbard band (UHB) as

$$G_{22}^{(0)}(k, E) = \langle\langle X_k^{-\sigma,2} | X_k^{2,-\sigma} \rangle\rangle_E = \frac{F_{-\sigma,2}}{E - \varepsilon_1 - U - F_{-\sigma,2}t(k)}. \tag{1.57}$$

From Eqs. (1.56) and (1.57), it is obvious that operators $X^{0,\sigma}$ and $X^{-\sigma,2}$ describe the quasiparticle of Fermi type in LHB and UHB, respectively.

As will be shown in the Chapter 5, the Hubbard 1 decoupling scheme is equivalent to the Hartree–Fock approximation in the X-operator diagram technique. Half-field Hubbard model with $n_e = 1$ in the limit $U \gg W$ describes the Mott–Hubbard insulator with empty UHB and occupied LHB. When there is doping by electrons ($n_e = 1 + x, x > 0$) or by holes ($n_e = 1 - x, x > 0$), the Fermi level moves into the UHB or LHB, correspondingly. At very small x the question whether the extra carriers are itinerant or localized cannot be solved in the Hubbard 1 scheme because the carrier scattering on the spin and charge fluctuations may result in its localization below the mobility edge. Above the mobility edge the carriers in the Hubbard bands correspond to the metal state. There is a Fermi surface determined as usually by the conditions $E_1(k) = 0$ for LHB and $E_2(k) = 0$ for UHB, where the energies of the Hubbard bands are equal to

$$E_1(k) = \varepsilon + F_{0\sigma}t(k) - \mu,$$
$$E_2(k) = \varepsilon + U + F_{-\sigma 2}t(k) - \mu. \tag{1.58}$$

In the literature there is a widespread opinion that the Fermi surface in the Hubbard bands is too large and violates the Luttinger theorem (Fig. 1.4). It is a wrong opinion obtained when the spectral weight effect is neglected. Indeed, the Fermi surface for LHB in Fig. 1.4 is larger than for free electrons, but every quantum cell in the k-space contains $F_{0\sigma}$ ($F_{-\sigma 2}$) electrons for LHB (UHB) instead of 1 electron for free particles. The spectral weight redistribution, as can be seen from Eqs. (1.56) and (1.57), results in the modified equation for the chemical potential

$$n_e = \frac{1}{N} \sum_{k\sigma} \left(F_{0\sigma} f_F(E_1(k)) + F_{-\sigma 2} f_F(E_2(k)) \right),$$
$$f_F(E) = [\exp(E/T) + 1]^{-1} \tag{1.59}$$

that provides the validity of the Luttinger theorem for the Hubbard bands. In other words, the Fermi surface of the Hubbard quasiparticles is more compressible, not so dense as for the free electrons.

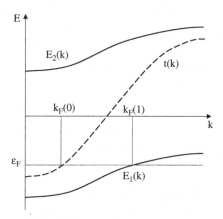

Fig. 1.4 The free electron band $t(k)$, the LHB $E_1(k)$, and the UHB $E_2(k)$. The same Fermi level results in $k_F^{(0)}$ for free electrons and in $k_F^{(1)}$ for LHB.

In the limit $U = \infty$ in the paramagnetic phase at $T = 0$, we have

$$\langle X^{22} \rangle = 0,$$
$$F_{-\sigma 2} = \langle X^{-\sigma,-\sigma} \rangle = n_e/2,$$
$$F_{0\sigma} = 1 - F_{-\sigma 2} = 1 - n_e/2.$$

The square Fermi surface determined by Eq. (1.10) takes place for $n_e = 2/3$, as can be seen from Eq. (1.59). Thus, strongly correlated quasiparticle and free electron may have the Fermi surface of similar shape, when probed at different electron concentration.

Beyond the Hubbard 1 approximation one can include the quasiparticle high order scattering processes and the interaction of the Fermi type quasiparticle with the Bose type spin and charge fluctuations. Nevertheless, even in simple Eqs. (1.56) and (1.57), one can see the following SEC effects:

(a) The correlation narrowing of the quasiparticle bandwidth that is determined self-consistently by the filling factors. The effective bandwidth becomes dependent on temperature, magnetic field and electron concentration.
(b) The spectral weight decreasing for each Hubbard subband caused by the filling factor in the numerator of Green's functions (1.56, 1.57). Due to the sum rule, only the total number of states for both LHB and UHB is 1 for each spin projection and 2 for both spin projections. In other words the Fermi type quasiparticles in the X-operator representation have both the electrical charge ($n_p - n_q = \pm 1$) and the spin $S = 1/2$. The principal difference from the free particles is the reduced spectral weight that is calculated self-consistently and depends on the temperature, magnetic field and electron concentration. Due to the sum rule the spectral weight of one quasiparticle depends on the spectral weight of the other quasiparticle.

Since the Hubbard 1 approximation contains these essential SEC effects it is more convenient to use it as a zero approximation for the description of the SEC system then the weak coupling approach which starts from the band limit. Several methods are known in the literature on how to proceed beyond the Hubbard 1 approximation and the diagram technique (Chapter 5) is one of them. In the limit $t \ll U$, the Hubbard 1 solution is quite reasonable and it will be used later (Chapters 3 and 8) to analyse the quasiparticle band structure of the Mott–Hubbard insulators such as the transition metal oxides.

1.10 Study of the Electronic Structure by the Angle Resolved Photoemission Spectroscopy

Angle resolved photoemission spectroscopy (ARPES) is a powerful technique to investigate the occupied electronic states below the Fermi level in solids. The incident photons illuminate the surface of a single crystalline sample and then a photoelectron is ejected from the surface. The kinetic energy and the angle of the outgoing electron are measured, and the energy and momentum distributions of the occupied state are constructed (Hüfner 1995). The recent development in this field, stimulated mainly by the physics of high-T_C superconductors, has led to significant improvement of angle and energy resolutions (Damascelli, Lu and Shen 2001). The Fermi surface in hole-doped superconductors Bi2212, YBCO, and electron-doped $Nd_{2-x}Ce_xCuO_4$, the electron band structure in the undoped insulator $Sr_2CuO_2Cl_2$, the doping dependence in LSCO, the superconducting gap, and the pseudo-gap phenomenon in the normal state have all been studied by ARPES (see the review of ARPES data and their theoretical discussions by Tohyama and Maekawa (2000)).

The two-dimensional nature of the CuO_2 plane in high T_C cuprates is favorable to ARPES in determining the in-plane electronic states below the Fermi level, since the measured angle θ determines only the component of momentum $p_\parallel$ parallel to the surface of sample. Nevertheless a lot of ARPES studies of different materials, not only cuprates, have been carried out recently. We note that to obtain reliable data it is necessary to be careful of the quality of the surface, since ARPES is a surface sensitive technique due to short escape depth (~ 10 Å) of the outgoing electron.

We will describe ARPES data following the review by Tohyama and Maekawa (2000).

To understand the ARPES data one has to take into account a complicated process involving initial state and final state interactions, as well as interactions between the outgoing electron and the surface, and so on. An effort to include all of the effects has been made based on the so-called one-step model in which the absorption of the photon and the emission of photoelectron are treated as a single event (Bansil and Lindroos 1998). However, since the method relies on band theory within the LDA approximation that failed for SEC systems, we do not know how reliable these results are.

Instead of the one-step calculation, we adopt the sudden approximation (Hedin and Lundquist 1996; Hüfner 1995) where final state interactions are neglected. The ARPES intensity is then given by

$$I(k,\omega) = I_0(k)f(\omega)A(k,\omega) \qquad (1.60)$$

where k is the momentum parallel to the surface, ω is the excitation energy measured from the Fermi level, and $I_0(k)$ is the squared matrix element that mainly comes from the condition of the dipole transition and determines the dependence of $I(k,\omega)$ on the polarization of the incident photon. The Fermi function is

$$f(\omega) = [\exp(\beta\omega) + 1]^{-1}, \quad \beta = 1/k_B T.$$

The spectral function $A(k,\omega)$ consists of the electron-removal A_- and electron-addition A_+ terms:

$$A(k,\omega) = A_-(k,\omega) + A_+(k,\omega).$$

They are given by

$$A_\pm(k,\omega) = Z^{-1} \sum_{nm\sigma} \exp(-\beta E_m^N)|\langle \psi_n^{N\pm 1}|a_{k\sigma}|\psi_m^N\rangle|^2 \delta(\omega \pm E_n^{N\pm 1} \mp E_m^N)$$

(1.61)

where $a_{k\sigma} = c_{k\sigma}^+$ and $c_{k\sigma}$ for A_+ and A_-, respectively; $c_{k\sigma}^+(c_{k\sigma})$ is the creation (annihilation) operator of an electron with momentum k and spin σ; ψ_m^N is the wavefunction of the m-th eigenstate with energy E_m^N in the N-electron system. Here, $Z = \sum_m \exp(-\beta E_m^N)$ is the partition function. We note that $f(\omega)A(k,\omega) = A_-(k,\omega)$ providing that ARPES investigates the occupied states below the Fermi level. Similarly, the inverse photoemission spectrum ARIPES is determined by the $A_+(k,\omega)$ function and investigates the unoccupied states above the Fermi level.

As was introduced above in Sec. 1.6, the spectral function $A(k,\omega)$ is proportional to the imaginary part of the single electron Green's function $G(k,\omega)$. Thus *ab initio* calculation of ARPES is possible for the conventional Fermi liquid systems where LDA results in the correct electronic spectrum. For the SEC systems, no *ab initio* Green's function calculations are available; the theoretical study of the ARPES data has to be restricted to some simplified models like the Hubbard or the t–J models. The exact diagonalization and the quantum Monte-Carlo studies of the finite clusters are very useful, because these approaches deal with the multielectron states $|\psi_m^N\rangle$ explicitly (Tohyama and Maekawa 2000). For the infinite systems several perturbation theories have been developed. The dynamical cluster approximation to the ARPES in the two-dimensional Hubbard model has been used recently by Maier, Pruschke, and Jarrell (2000). Later in this book we will introduce the generalized tight binding method (Chapter 3, and it will be used in the Chapter 8) to calculate the ARPES and its temperature and concentration dependence in the hole doped cuprates.

Chapter 2

Multielectron Models in X-Operator Representation

In this chapter several popular models used for SEC systems are formulated in the X-operator representation. Since local magnetic moments (LMM) in magnetic insulators and semiconductors are formed by the multielectron configurations of $d(f)$-ions, we consider also isotropic and single-ion anisotropic Heisenberg exchange models.

2.1 The Isotropic Heisenberg Model

The Hamiltonian of LMM with spin S coupled by the interatomic exchange interaction I_{fg} in the external magnetic field H oriented along z-axis is given by

$$H = -\frac{1}{2}\sum_{fg} I_{fg} \mathbf{S}_f \cdot \mathbf{S}_g - g\mu_B H \sum_f S_f^z. \qquad (2.1)$$

We define the X-operators $X_f^{M_1 M_2} = |M_1\rangle\langle M_2|$ using the set of eigenstates of S_f^z operator

$$S_f^z |M\rangle = M|M\rangle, \quad M = -S, -S+1, \ldots, S.$$

Using Eq. (1.37), we can write the X-representation of the spin operators

$$S_f^z = \sum_{M=-S}^{S} M X_f^{M,M}, \quad S_f^+ = \sum_{M=-S}^{S} \nu(M) X_f^{M+1,M}, \quad S_f^- = (S_f^+)^*, \qquad (2.2)$$

where $\nu(M) = ((S-M)(S+M+1))^{1/2}$. For the space uniform ferromagnetic phase, it is convenient to add and subtract the mean field $I_0 \langle S^z \rangle$ from the

Hamiltonian (2.1), where

$$I_0 \equiv I(q=0), \quad I(q) = \sum_{f-g} I_{fg} \exp(i q \, (\bm{R}_f - \bm{R}_g)). \tag{2.3}$$

Then we obtain the Hamiltonian (2.1):

$$H = H_0 + H_{\text{int}} + \frac{1}{2} N I_0 \langle S^z \rangle^2, \tag{2.4}$$

$$H_0 = \sum_{fM} E_M X_f^{M,M}, \quad E_M = -(g \mu_B H + I_0 \langle S^z \rangle) M, \tag{2.5}$$

$$H_{\text{int}} = -\frac{1}{2} \sum_{fg} \left\{ \sum_{\alpha\beta} I_{fg}^{\alpha\beta} X_f^{\alpha} X_g^{-\beta} + \sum_{M_1,M_2} I_{fg}^{M_1,M_2} \left(X_f^{M_1,M_1} - \langle X_f^{M_1,M_1} \rangle \right) \right.$$
$$\left. \times \left(X_g^{M_2,M_2} - \langle X_g^{M_2,M_2} \rangle \right) \right\}. \tag{2.6}$$

Here the interaction matrix has a multiplicative form

$$I_{fg}^{\alpha\beta} = I_{fg} \nu(\alpha) \nu(\beta), \quad I_{f,g}^{M_1,M_2} = I_{fg} M_1 M_2,$$
$$\nu(\alpha) \equiv \nu(M) = [(S-M)(S+M+1)]^{1/2}, \quad \alpha = \alpha(M+1, M), \tag{2.7}$$

while for all root vectors $\alpha \neq \alpha(M+1, M)$, $\nu(\alpha) = 0$. Certainty, the example of the isotropic Heisenberg model is rather trivial, as there is more simple and elaborated diagram technique for the spin operators (Vaks et al 1967a,b; Izyumov et al 1974). Nevertheless, the X-operator formulation of the Heisenberg model has its own advantage. It is possible to obtain both the spin wave excitation spectra and population numbers n_M of all single ion states $|M\rangle, n_M = \langle X^{M,M} \rangle$ simultaneously by calculating the Green functions $\langle\langle X_f^{\alpha} | X_g^{\beta} \rangle\rangle$. It simplifies the derivation of the self-consistent equation for the order parameter $\langle S^z \rangle$ for the arbitrary value of spin S. This simplification allowed Val'kov and Ovchinnikov (1982) to get the compact expression for the sublattice magnetization of the isotropic Heisenberg antiferromagnetic with an arbitrary value of the spin S with regard to zero spin fluctuations. At $T = 0$ it is given by

$$\langle S^z \rangle_{T=0} = S \, B_S \left[S \ln \frac{1 + P_S(0)}{P_S(0)} \right], \tag{2.8}$$

where $B_S(x)$ is the known Brillouin function and the function $P_S(T)$ is given by

$$P_S(0) = \frac{1}{2N}\sum_q \{(1-\gamma_q^2)^{-1/2} - 1\}, \quad \gamma(q) = \frac{1}{|I|z}\sum_h I(h)\exp(i\mathbf{q}\cdot\mathbf{h}). \tag{2.9}$$

Expression (2.8) allows a smooth transition from the quantum $S = 1/2$ antiferromagnet with $(S - \langle S^z \rangle)/S = 0.0675$ up to the quasiclassical limit $S \gg 1$ where $(S - \langle S^z \rangle)/S = 0.078$. The more important advantage of the X-operator representation for magnetic insulator appears in the case of a strong single-ion anisotropy (SA) that is considered below.

2.2 The Heisenberg Model with Single-Ion Anisotropy

Even a simplest single-ion anisotropy contribution to the Hamiltonian

$$H_A = -D\sum_f (S_f^z)^2 \tag{2.10}$$

results in serious difficulties for the conventional theory of magnetic insulator dealing with spin operators. At small D the SA may be considered using a perturbation method and it causes magnon-magnon interaction. In many systems, SA is not small, for example in the rare earth and actinide ions. In such cases, it is desirable to include H_A in the zeroth order part of the Hamiltonian H_0. For the SA, this results in the nonequidistant energy spectrum of an ion that prohibits the application of the conventional spin operator method. Nevertheless for X-operators it is not a problem. Not only the simplest form of H_A (2.10) but rather the general form of SA

$$H_A = -\sum_f \varphi(S_f^z), \tag{2.11}$$

with an arbitrary analytical function $\varphi(S_f^z)$ may be treated on an equal footing (Val'kov and Ovchinnikov 1983). In the X-operators representation all complicated dependence $\varphi(S_f^z)$ transforms into c-number function $\varphi(M)$, because

$$H_A = -\sum_f \sum_{M=-S}^{S} \varphi(M) X_f^{M,M}. \tag{2.12}$$

In the X-operators representation, the arbitrary SA is included in H_0 and just renormalizes ion's energy levels in Eq. (2.5):

$$E_M = -(g\mu_B H + I_0 \langle S^z \rangle)M - \varphi(M). \tag{2.13}$$

With this renormalization the SA Heisenberg model and isotropic Heisenberg model have the same operator structure (2.4) and both models can be studied perturbatively on an equal footing.

2.3 The Atomic Representation for s–$d(f)$ Exchange Model

This model is very popular in magnetic semiconductors and metals containing both LMM with spin S and some itinerant electrons (we call them s-electrons but they may arise from p- or d-electron orbital as well). The LMM is forming by d^n (f^n) configuration. The number of localized $d\,(f)$ electrons and the value of spin S in the model are fixed, it means that only the spin but not the charge fluctuations are considered. Here we write more a general form of the s–d model Hamiltonian, which retain the main ideas of the initial model.

The Hamiltonian of LMM with spin S may be written as follows

$$H_d = -\sum_f \sum_{M=-S}^{S} (E_0 - n\mu) X_f^{M,M}, \tag{2.14}$$

where $X^{M,M} = |M\rangle\langle M|$ and the $(2S+1)$-fold spin degeneracy in the paramagnetic phase is shown explicitly. The itinerant part of the Hamiltonian is given by the Hubbard Hamiltonian

$$H_s = \sum_{f\sigma}(\varepsilon - \mu)n_{f\sigma} + \frac{U}{2}\sum_{f\sigma} n_{f\sigma}n_{f,-\sigma} + \sum_{f,f',\sigma}(t_{ff'} a_{f\sigma}^+ a_{f'\sigma} + \text{h.c.}). \tag{2.15}$$

The s–$d\,(f)$ exchange interaction is supposed to be local

$$H_{s-d} = -J \sum_f \mathbf{S}_f \cdot \boldsymbol{\sigma}_f, \tag{2.16}$$

with $\hat{S}_f$ and $\hat{\sigma}_f$ being the spin operators for localized and itinerant electrons correspondingly. The Zeeman energy in the external magnetic field is given by

$$H_z = -h\sum_f \left(S_f^z + \sigma_f^z\right). \quad (2.17)$$

The total Hamiltonian of the s–d(f) exchange model is

$$H = H_d + H_s + H_{s-d} + H_z, \quad (2.18)$$

and we can present as the sum of single-site and two-site terms

$$H = H_0 + H_1, \quad H_0 = \sum_f H_f, \quad H_1 = \sum_{f,f'} H_{ff'}. \quad (2.19)$$

We start with the exact diagonalization of the local Hamiltonian H_f. When $J = 0$ the $d^n(f^n)$-ion has $(2S+1)$ eigenstates $|M\rangle$ while the itinerant electrons have 4 local eigenstates $|0\rangle$, $|\sigma\rangle$, $|2\rangle$, and the total number of the local states is $4(2S+1)$. At $J \neq 0$ the eigenstates of H_{sd} have been obtained by Nagaev (1983). The eigenstates of H_f are denoted by the electron number $n_e = n + \rho$ ($\rho = 0, 1, 2$), spin and spin projection and can be written in the following form

$$d^n : |S, M, 0\rangle \equiv |M, 0\rangle = |M\rangle|0\rangle,$$
$$s\,d^n : |S+1/2, M, 1\rangle \equiv |M, +\rangle = u_M|M-1/2\rangle\|\uparrow\rangle + v_M|M+1/2\rangle|\downarrow\rangle,$$
$$|S-1/2, M, 1\rangle \equiv |M, -\rangle = v_M|M-1/2\rangle\|\uparrow\rangle - u_M|M+1/2\rangle|\downarrow\rangle,$$
$$s^2 d^n : |S, M, 2\rangle \equiv |M, 2\rangle = |M\rangle|2\rangle \quad (2.20)$$

with u, v-coefficients given by

$$u_M^2 = \frac{S+M+1/2}{2S+1}, \quad v_M^2 = \frac{S-M+1/2}{2S+1}, \quad u_M^2 + v_M^2 = 1.$$

Using the exact eigenstates $|f,\gamma\rangle$ of H_f (2.20), we introduce new Hubbard operators $X_f^{\gamma\gamma'} = |\gamma\rangle\langle\gamma'|$ and in these operators representation H_f becomes diagonal

$$H_f = \sum_{M=-S}^{S} E_{M,0} X_f^{M,0;M,0} + \sum_{M=-S-1/2}^{S+1/2} E_{M,+} X_f^{M,+;M,+}$$
$$+ \sum_{M=-S+1/2}^{S-1/2} E_{M,-} X_f^{M,-;M,-} + \sum_{M=-S}^{S} E_{M,2} X_f^{M,2;M,2}. \quad (2.21)$$

Equation (2.21) describes the superposition of different multielectron configurations: d^n with spin S, sd^n in both high spin $S + 1/2$ and low spin

$S-1/2$ states and s^2d^n with spin S. The energies of these configurations are equal to

$$\begin{aligned} E_{M,0} &= E - n\mu - hM, \\ E_{M,+} &= E_0 + \varepsilon - JS/2 - hM - (n+1)\mu, \\ E_{M,-} &= E_0 + \varepsilon + J(S+1)/2 - hM - (n+1)\mu, \\ E_{M,2} &= E_0 + 2\varepsilon + U - hM - (n+2)\mu. \end{aligned} \quad (2.22)$$

The electron creation (annihilation) operator in the X-operator representation is given by (here $\sigma = \pm 1/2$):

$$a_{f\sigma}^+ = \sum_{M=-S}^{S} \{ 2\sigma u_{M+\sigma} X_f^{M+\sigma,\sigma;M,0} + v_{M+\sigma} X_f^{M+\sigma,-\sigma;M,0} \\ - 2\sigma u_{M-\sigma} X_f^{M,2;M-\sigma,-\sigma} + v_{M-\sigma} X_f^{M,2;M-\sigma,\sigma} \}. \quad (2.23)$$

Using the root vectors the last expression may be simplified as

$$a_{f\sigma}^+ = \sum_\alpha \gamma_\sigma(\alpha) X_f^\alpha, \quad a_{f\sigma} = \sum_\alpha \gamma_\sigma(\alpha) X_f^{-\alpha}, \quad (2.24)$$

where the matrix elements $\gamma(\alpha)$ are equal to

$$\gamma(\alpha) = \begin{cases} v_{M+\sigma} & \alpha = \alpha_1(|M+\sigma,-\sigma\rangle, |M,0\rangle) \\ 2\sigma u_{M+\sigma} & \alpha = \alpha_2(|M+\sigma,\sigma\rangle, |M,0\rangle) \\ v_{M-\sigma} & \alpha = \alpha_3(|M,2\rangle, |M-\sigma,\sigma\rangle) \\ -2\sigma u_{M-\sigma} & \alpha = \alpha_4(|M,2\rangle, |M-\sigma,-\sigma\rangle) \\ 0 & \alpha \neq \alpha_1, \alpha_2, \alpha_3, \alpha_4 \end{cases} \quad (2.25)$$

The total spin operator $\boldsymbol{S}_{\text{tot}} = \boldsymbol{S} + \boldsymbol{\sigma}$ has the following components:

$$S_f^+ = \sum_{M=-S}^{S} \nu_S(M) X_f^{M+1,0;M,0} + \sum_{M=-S-1/2}^{S+1/2} \nu_{S+1/2}(M) X_f^{M+1,+;M,+} \\ + \sum_{M=-S+1/2}^{S-1/2} \nu_{S-1/2}(M) X_f^{M+1,-;M,-} + \sum_{M=-S}^{S} \nu_S(M) X_f^{M+1,2;M,2}, \quad (2.26)$$

$$S_f^z = \sum_{M=-S}^{S} M X_f^{M,0;M,0} + \sum_{M=-S-1/2}^{S+1/2} M X_f^{M,+;M,+} \\ + \sum_{M=-S+1/2}^{S-1/2} M X_f^{M,-;M,-} \sum_{M=-S}^{S} M X_f^{M,2;M,2}. \quad (2.27)$$

The interatomic part of the Hamiltonian is given by

$$H_1 = \sum_{ff'} t_{ff'} \sum_{\alpha\beta} \gamma_\alpha \gamma_\beta X_f^\alpha X_{f'}^{-\beta}. \tag{2.28}$$

Thus the total Hamiltonian of the s–d-exchange model given by the sum of Eqs. (2.21) and (2.28) has exactly the same operator structure as the Hubbard model Hamiltonian (Erukhimov and Ovchinnikov 1984). The only difference is the dimension of local basis set, here it is equal to $4(2S+1)$. The operator equivalence of s–d and Hubbard model allows one to apply the perturbation approach developed for the Hubbard model like the diagram technique for the X-operators. Some exact results obtained for the Hubbard model may also be applied to the s–d model such as the Nagaoka theorem. For the s–d model with $U = \infty$ and $J = +\infty$ the same procedure as that used by Nagaoka (1966) results in the exact ferromagnetic ground state in the following situations:

(1) number of electrons $N_e = Nn+1$ (one extra electron on the background of d^n-ions),
(2) $N_e = N(n+1) \pm 1$ (one extra electron or hole on the background of high spin $S + 1/2$), and
(3) $N_e = N(n+2) - 1$ (one hole on the background of $s^2 d^n$-configurations with spin S).

2.4 The Periodic Anderson Model

In this model the localized $d(f)$ electron may fluctuate between the two multielectron configurations of d^n with spin S_n and d^{n+1} with spin S_{n+1}. While the X-operator representation is suitable for an arbitrary spin value S we restrict ourselves here to the model with $S_n = 0$ and $S_{n+1} = 1/2$, $S_{n+1}^z = \pm\sigma$, which is usually discussed. The d-electron term in the Hamiltonian is given by

$$H_d = \sum_f \left\{ (E_n - n\mu)X_f^{0,0} + \sum_\sigma (E_{n+1} - (n+1)\mu)X_f^{\sigma,\sigma} \right\}. \tag{2.29}$$

The band electrons are treated in the conventional tight-binding scheme

$$H_c = \sum_{f\sigma} (\varepsilon - \mu) n_{f\sigma} + \sum_{ff'\sigma} (t_{ff'} a_{f\sigma}^+ a_{f'\sigma} + \text{h.c.}). \tag{2.30}$$

The hybridization of itinerant and localized electrons may be written as follows

$$H_{cd} = V \sum_{f\sigma} \left(X_f^{\sigma,0} a_{f\sigma} + a_{f\sigma}^+ X_f^{0,\sigma} \right). \tag{2.31}$$

The total Hamiltonian of the periodic Anderson model is given by

$$H = H_d + H_c + H_{cd}. \tag{2.32}$$

This model has been intensively studied in the mixed valence and heavy fermion problems. Again we write the Hamiltonian as the sum of single-site and two-site terms. The exact diagonalization of the single-site Hamiltonian H_f has been done by Foglio and Falikov (1979). It is easy to see that the operators of electron numbers $\hat{N}_f$ and the spin projection $\hat{S}_f^z$ at a given site

$$\begin{aligned}\hat{N}_f &= \sum_\sigma a_{f\sigma}^+ a_{f\sigma} + n X_f^{0,0} + (n+1) \sum_\sigma X_f^{\sigma,\sigma}, \\ \hat{S}_f^z &= \sum_\sigma \sigma \left(a_{f\sigma}^+ a_{f\sigma} + X_f^{\sigma,\sigma} \right)\end{aligned} \tag{2.33}$$

commute with H_f, so the eigenvectors of H_f are denoted by electron numbers $n_e = n + \rho$ ($\rho = 0, 1, 2, 3$), spin value $S = 0, 1/2, 1$ and S^z. When $V = 0$ there are 12 different states with even valence:

$$d^n : |G, 0\rangle \equiv |G\rangle|0\rangle,$$
$$d^{n+1} : |+, 0\rangle = |+\rangle|0\rangle, |-, 0\rangle = |-\rangle|0\rangle,$$
$$cd^n : |G, \uparrow\rangle = |G\rangle|\uparrow\rangle, |G, \downarrow\rangle = |G\rangle|\downarrow\rangle,$$
$$cd^{n+1} : |+, \uparrow\rangle = |+\rangle|\uparrow\rangle, |+, \downarrow\rangle = |+\rangle|\uparrow\rangle, |-, \uparrow\rangle = |-\rangle|\uparrow\rangle, |-, \downarrow\rangle = |-\rangle|\downarrow\rangle,$$
$$c^2 d^n : |G, 2\rangle = |G\rangle|2\rangle,$$
$$c^2 d^{n+1} : |+, 2\rangle = |+\rangle|2\rangle, |-, 2\rangle = |-\rangle|2\rangle,$$

where $|G\rangle$ is the singlet ground state of d^n-ion, $|\pm\rangle$ are the states of d^{n+1}-ion with $S = 1/2$ and spin projection σ, $|0\rangle, |\uparrow\rangle, |\downarrow\rangle$ and $|2\rangle$ are the local states of itinerant electrons in the nondegenerate band of the s-type. The hybridization mixes c and d states but the total number of electrons and the spin projection are conserving, so the 12 × 12 matrix of H_f can be subdivided into several blocks and easily diagonalized. The 12 eigenvectors $|p\rangle$ are the following:

- 3 singlets ($S = 0$),
- 3 doublets ($S = 1/2$), and
- 1 triplet ($S = 1$),

with energies and wavefunctions given by

(1) $\rho = 0$, $S = 0$, $|0\rangle = |G, 0\rangle$, $E_0 = E_n - n\mu$. (2.34)

(2) $\rho = 1$, $S = 1/2$, $|1, \sigma, +\rangle = u_1|G, \sigma\rangle + v_1|\sigma, 0\rangle$,

$|1, \sigma, -\rangle = -v_1|G, \sigma\rangle + u_1|\sigma, 0\rangle$, (2.35)

$$E_{1,\sigma}^{(\pm)} = E_0 + \frac{1}{2}(\varepsilon + \omega \pm v_1) - \mu,$$

$$\omega = E_{n+1} - E_n, \quad v_1^2 = (\varepsilon - \omega)^2 + 4V^2.$$

(3) $\rho = 2$, $S = 1$, $S^z = 2\sigma$, $|T, 2\sigma\rangle = |\sigma, \sigma\rangle$

$$S^z = 0, \quad |T, 0\rangle = \frac{(|+, \downarrow\rangle + |-, \uparrow\rangle)}{\sqrt{2}},$$

$$E_{2,T} = E_0 + \varepsilon + \omega - 2\mu,$$

$S = 0, |S, +\rangle = u_2|G, 2\rangle + v_2(|+, \downarrow\rangle - |-, \uparrow\rangle)/\sqrt{2}$,

$|S, -\rangle = -v_2|G, 2\rangle + u_2(|+, \downarrow\rangle - |-, \uparrow\rangle)/\sqrt{2}$,

$$E_S^{(\pm)} = E_0 + \varepsilon + \frac{1}{2}(\varepsilon + \omega \pm v_2) - 2\mu,$$

$$v_2^2 = (\varepsilon - \omega)^2 + 8V^2. \quad (2.36)$$

(4) $\rho = 3$, $S = 1/2$, $|\sigma, 2\rangle = |\sigma\rangle|2\rangle$, $E_{3,\sigma} = E_0 + 2\varepsilon + \omega - 3\mu$. (2.37)

Here $u_i^2 = \frac{1}{2}(1 + (\varepsilon - \omega)/\nu_i)$ and $v_i^2 = 1 - u_i^2$ ($i = 1, 2$). A straightforward calculation of the matrix elements of the Fermi creation operator $\langle p|a_{f\sigma}^+|q\rangle$ with the local basis (2.34–2.37) gives a nonzero contribution to the 14 root vectors $\boldsymbol{\alpha}_m$

$$a_{f\sigma}^+ = \sum_{m=1}^{14} \gamma_m X_f^{\alpha_m}. \quad (2.38)$$

Matrix elements γ_m and root vectors $\boldsymbol{\alpha}_m$ are presented in the Table 2.1. Finally the periodic Anderson model in the X-operator representation can be written as the generalized Hubbard model with 12 local states

$$H = H_0 + H_1, \quad (2.39)$$

$$H_0 = \sum_f \sum_{p=1}^{12} E_p X_f^{pp},$$

$$H_1 = \sum_{f,g} t_{fg} \sum_{m,n=1}^{14} \gamma_m \gamma_n^* X_f^{\alpha_m} X_g^{-\alpha_n}.$$

Table 2.1 The nonzero matrix elements of the Fermi operator in the X-operator representation for the Periodic Anderson model. Here $\alpha_m = (p, q)_m$

m	1	2	3	4	5	6	7
α_m	$(1, \sigma, +; 0)$	$(1, \sigma, -; 0)$	$(T, 2\sigma; 1+; 0)$	$(T, 0; 1, -\sigma, +)$	$(S, +; 1, \sigma, +)$	$(S, -; 1, \sigma, +)$	$(T, 2\sigma; 1\sigma, -)$
γ_m	u_1	$-v_1$	$-v_1$	$\dfrac{-v_1}{\sqrt{2}}$	$\dfrac{2\sigma(u_1 u_2 + v_1 v_2)}{\sqrt{2}}$	$\dfrac{-2\sigma(u_1 v_2 - v_1 u_2)}{\sqrt{2}}$	$-u_1$

m	8	9	10	11	12	13	14
α_m	$(T, 0; 1, -\sigma, -)$	$(S, +; 1, -\sigma, -)$	$(S, -; 1, -\sigma, -)$	$(-\sigma, 2; T, 0)$	$(\sigma, 2; T, 0)$	$(\sigma, 2; S, +)$	$(\sigma, 2; S, -)$
γ_m	$\dfrac{u_1}{\sqrt{2}}$	$\dfrac{2\sigma(v_1 u_2 + u_1 v_2)}{\sqrt{2}}$	$\dfrac{2\sigma(v_1 v_2 + u_1 u_2)}{\sqrt{2}}$	-2σ	$\dfrac{-1}{\sqrt{2}}$	$\dfrac{-v_2}{\sqrt{2}}$	$\dfrac{-u_2}{\sqrt{2}}$

Let us consider the particular case $\omega - \varepsilon > 0$ when the d^n configuration is more stable then the d^{n+1} configuration. In this case the low energy physics is described by 4 local states

$$\begin{aligned} |0\rangle & \quad (\rho = 0), \\ |1, \sigma, -\rangle & \quad (\rho = 1), \\ |S, -\rangle & \quad (\rho = 2) \end{aligned} \quad (2.40)$$

instead of the full basis. Neglecting the exponentially small occupation of all the other states and taking into account only 4 states (2.40) we can see that the local basis (2.40) looks like the basis for the usual Hubbard model with the only difference that the states $|1, \sigma, -\rangle$ and $|S, -\rangle$ are mixed valence states with one and two electrons correspondingly.

The operator equivalence of periodic Anderson model and Hubbard model was used by Ovchinnikov and Sandalov (1983) to calculate the quasiparticle self-energy by the X-operator diagram technique. They have found that the Fermi liquid behavior takes place in the strong correlation limit at very low temperatures and small quasiparticle band filling (when ρ is close to an even value).

2.5 The Multiband Model of Transition Metal Oxides

The transition metal oxides are typical examples of SEC systems. The chemical bonding is mainly the ionic one. The covalent contribution is much smaller but very important for the electronic structure formation. We will consider lattices with a metal ion surrounded by the oxygen octahedron (it may be distorted as well). Besides binary oxides of the NiO type the other examples are given by cuprates La_2CuO_4, nickelates La_2NiO_4, manganites La_2MnO_4 and others. Two opposing trends are characteristic of their electronic structure: strong p–d hybridization of the oxygen and cation orbitals that can be easily diagonalized in k-space, and SEC that are more simply treated in the local approach. To incorporate both of them Ovchinnikov and Sandalov (1989) have proposed the cluster perturbation approach when the total Hamiltonian is divided into the sum of the single-cluster (it may be a unit cell) and the intracluster terms, with the intracluster part H_f being diagonalized exactly. The thus obtained multielectron molecular orbitals of the cluster are used to construct the Hubbard operators X_f for the cluster. At the next step the intracluster hopping and interaction are written in the X-operator representation resulting in the generalized Hubbard model.

We consider the realistic p–d model of oxides with the Hamiltonian

$$H = H_d + H_p + H_{pd} + H_{pp}, \qquad (2.41)$$

$$H_d = \sum_r H_d(r),$$

$$H_d(r) = \sum_{\lambda\sigma}\left[(\varepsilon_{d\lambda} - \mu)d^+_{r\lambda\sigma}d_{r\lambda\sigma} + \frac{1}{2}U_d n^\sigma_{r\lambda} n^{-\sigma}_{r\lambda}\right]$$

$$+ \sum_{\sigma\sigma'}\left(V_d n^\sigma_{r1} n^{\sigma'}_{r2} - J_d d^+_{r1\sigma}d_{r1\sigma'}d^+_{r2\sigma'}d_{r2\sigma}\right),$$

$$H_p = \sum_i H_p(i),$$

$$H_p(i) = \sum_{\alpha\sigma}\left[(\varepsilon_{p\alpha} - \mu)p^+_{i\alpha\sigma}p_{i\alpha\sigma} + \frac{1}{2}U_p n^\sigma_{i\alpha} n^{-\sigma}_{i\alpha}\right]$$

$$+ \left(V_p n^\sigma_{i1} n^{\sigma'}_{i2} - J_p p^+_{i1\sigma}p_{i1\sigma'}p^+_{i2\sigma'}p_{i2\sigma}\right),$$

$$H_{pd} = \sum_{\langle i,r \rangle} H_{pd}(i,r),$$

$$H_{pd}(i,r) = \sum_{\alpha\lambda\sigma\sigma'}\left(T_{\lambda\alpha}p^+_{i\alpha\sigma}d_{r\lambda\sigma} + \text{h.c.} + V_{\lambda\alpha} n^\sigma_{r\lambda} n^{\sigma'}_{i\alpha} - J_{\alpha\lambda}d^+_{r\lambda\sigma}d_{r\lambda\sigma'}p^+_{i\lambda\sigma'}p_{i\alpha\sigma}\right),$$

$$H_{pp} = \sum_{\langle i,j \rangle}\sum_{\alpha\beta\sigma}\left(t_{\alpha\beta}p^+_{i\alpha\sigma}p_{j\beta\sigma} + \text{h.c}\right).$$

Here H_d and H_p are the local energies of d- and p-electrons, H_{pd} describes the interatomic p–d hopping $T_{\lambda\alpha}$, Coulomb $V_{\lambda\alpha}$ and exchange $J_{\lambda\alpha}$ interactions. The last term H_{pp} contains the interatomic oxygen–oxygen hopping t_{pp}. We consider different d-orbitals λ and p-orbitals α, with regard to the crystal field splitting. For simplicity the intra-atomic Coulomb $U_d(U_p), V_d(V_p)$ and exchange $J_d(J_p)$ matrix elements are assumed to be the same for all orbitals, this simplification is not principal and the real matrix elements with orbital dependence can be trivially incorporated. All parameters of the Hamiltonian are assumed to be known either from *ab initio* calculations or from the experimental data. In many situations when the electron concentration is close to filled d-ion shells, it is more convenient to use the hole representation, for example in the case of cuprates.

In the cluster perturbation approach, we write the Hamiltonian

$$H = H_0 + H_1,$$
$$H_0 = \sum_f H_f, \quad H_1 = \sum_{fg} H_{fg}.$$

The exact diagonalization of H_f gives the eigenstates $|p\rangle$:

$$H_f|p\rangle = E_p|p\rangle,$$
$$H_f = \sum_m E_m X_f^{m,m}. \qquad (2.42)$$

With these states we define the cluster X-operators[1] $X_f^{pq} = |p\rangle\langle q|$ and express all Fermi operators d and p via the X-operators using Eq. (1.37). Then we can write all intracluster terms in the form

$$H_{fg} = T_{fg} \sum_{\alpha\beta} \gamma(\alpha)\gamma(\beta) X_f^\alpha X_g^{-\beta} + \sum_{mn} V_{fg}^{mn} X_f^{m,m} X_g^{n,n}. \qquad (2.43)$$

The intercluster p–d Coulomb interaction results in the second term in Eq. (2.43). The comparison of Eqs. (2.43) and (2.8) reveals the operator equivalence of such physically different models as the p–d model and

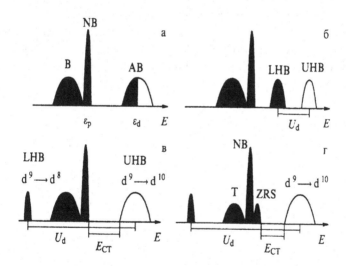

Fig. 2.1 Possible variants of the electronic structure in the three band p–d model: (a) metal, (b) Mott–Hubbard insulator with $U_d < \Delta$, (c) change transfer insulator with $U_d > \Delta$, (d) change transfer insulator with the splitted Zhang–Rice singlet and triplet states. Here the notations are: B-bonding, AB-antibonding, NB-non bonding, UHB and LHB are upper and lower Hubbard bands, ZRS-the Zhang–Rice singlet, T-triplet, E_{CT}-change transfer gap (Reprinted with permission from Horsch and Stephan 1993, in "Electronic Properties of High-T_c Superconductors", Eds. H. Kuzmany, M. Mehring and J. Fink (Springer Series in Solid State Science, **113**, 351 (1993)) © Springer).

---

[1] It is worthwhile mentioning that here the X-operator is not an atomic operator as considered above, but an operator for the cluster (unit cell).

the anisotropic Heisenberg one. The weak interatomic p–d Coulomb V_{pd} and exchange J_{pd} interactions (in comparison with the interatomic U_d, U_p terms) are often neglected, in this case only the non-diagonal X-operators given by the first term in Eq. (2.43) are presented and the p–d model becomes equivalent to the generalized multilevel Hubbard model. The calculations of the quasiparticle band structure of copper oxides in the framework of this model will be given below in Chapter 8.

The simplest variant of the p–d model is known as the 3-band p–d model (Emery (1987); Varma et al (1987)), where only one $d_{x^2-y^2}$ and one p-orbital (σ-bonding) are considered in the hole representation to model the electronic structure of cuprates. In the 3-band model the competition of local Coulomb correlations U_d and the charge transfer energy $\Delta = \varepsilon_p - \varepsilon_d$ determines possible variants of the electronic structure (Fig. 2.1). In the regime of SEC, when the d-electron band split in the LHB and the UHB, two insulator ground states are possible (Zaanen et al 1985). At $U_d > \Delta$ the Fermi level is in the gap between p-band and the Hubbard subband (charge transfer insulator). At $U_d < \Delta$ the Fermi level is in the gap between the UHB and the LHB (Mott–Hubbard insulator). For the 3-band p–d model the cluster (cell) perturbation method have been used in several papers (Lovtsov and Yushankhai 1991; Jefferson et al 1992; Belinicher et al 1996).

Chapter 3

General Approach to the Quasiparticle Description of Strongly Correlated Systems

The idea that an electron is a quasiparticle of the Landau Fermi liquid theory cannot be applied straightforwardly in the SEC systems because the SEC changes the ground state and may result in the metal–insulator transition with the Fermi surface disappearing at all. Nevertheless, the synthesis of the Landau and Hubbard ideas allows one to introduce the notion of a Fermi type quasiparticle for SEC systems as well as a Bose type quasiparticles (excitons, magnons etc.) as the excitations for the multielectron systems. In this chapter the definition of the Fermi and Bose quasiparticles will be given and the scheme to calculate their dispersion law which we call the generalized tight-binding (GTB) method is considered.

3.1 The Definition of Fermi- and Bose-Type Quasiparticles

There is a widespread opinion that quasiparticles in SEC systems obey a statistics other then the Fermi one, and that SEC separates the charge and spin degrees of freedom. There are several variants of such separations, in one case an electron is a combination of a neutral fermion with spin and a spinless charged boson, in the other cases the fermion is charged and spinless but the boson is neutral and has a spin. The very fact of such ambiguous representation induces some doubts in its correctness. Using the X-operator representation we show here that there is a rather natural way to define Fermi type single particle excitations with an electric charge e and spin $1/2$ but with a more complicated statistics, taking into account the constraints: due to SEC some sectors of the Hilbert space are pushed out of the low energy region.

We start from the consideration of local states where the intercell hopping is absent. The local eigenstates are shown in the Fig. 3.1 for the

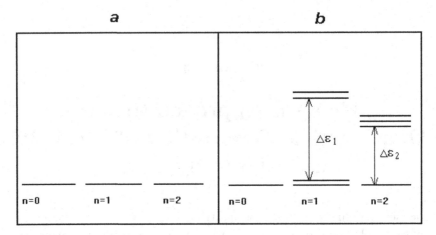

Fig. 3.1 Eigenstates with different number of electrons for the Hubbard (a) and multiband p–d (b) models. For p–d model high energy excited states with $n = 1, n = 2$ are not shown for simplisity.

Hubbard model (a) and the multiband p–d model (b), but only several low lying excited states with $n = 1$ and $n = 2$ are shown in the case (b). Following the Landau Fermi liquid theory we define the quasiparticle energy as a change of the N-electron energy when adding one more electron

$$\Omega_{pq} = \Omega_q(N+1) - \Omega_p(N), \qquad (3.1)$$

where $E_p(N)$ is the energy of p-th term for the N-electron system. Our difference from the Landau theory is that we use the definition (3.1) not for the whole system but only for the local cell where a quasiparticle appears. The hopping between cells and the resulting dispersion will be considered below.

For the Hubbard model there are two such quasiparticles

$$\Omega_- = \varepsilon_1 - \varepsilon_0 = \varepsilon_1, \quad \Omega_+ = \varepsilon_2 - \varepsilon_1 = \varepsilon_1 + U, \qquad (3.2)$$

corresponding to the LHB and UHB. In the multiband model the number of such quasiparticles may be rather large. Being constructed on the full orthogonal and normalized basis set of a unit cell, the X-operators provide the natural mathematical language of an adequate description of such quasiparticles. According to Eq. (1.37) any local operator can be presented as a linear combination of the X-operators, for an electron annihilation operator in the cell f with the band index λ, one has

$$a_{f\lambda\sigma} = \sum_{pq} |p\rangle\langle p|a_{f\lambda\sigma}|q\rangle\langle q| = \sum_{pq} \gamma_{\lambda\sigma}(pq) X_f^{pq}. \qquad (3.3)$$

Equation (3.3) shows clearly the difference in the Fermi type quasiparticle description in the single electron language and in the multielectron one. The operator $a_{f\lambda\sigma}$ decreases the number of electrons by one for all sectors of the Hilbert space simultaneously, while the X^{pq} operator describes the partial process of electron destruction in the $(N+1)$-electron configuration $|q\rangle$ with the final N-electron configuration $|p\rangle$. The matrix element $\langle p|a_{f\lambda\sigma}|q\rangle$ gives the probability of such process. If the problem of the exact diagonalization of cell Hamiltonian has been solved then the matrix elements are calculated straightforwardly. When SEC pushed out some sectors of the Hilbert space to the high-energy region it is natural to neglect the corresponding states $|p\rangle$ in the local basis as well as some partial processes of the electron destruction. In other words, working in the X-operator representation, it is easy to take into account the constraint induced by SEC where the single electron operator $a_{f\lambda\sigma}$ does not fit well. Therefore, the splitting of the single particle creation/annihilation process in the partial components is the first important step in the quasiparticle definition.

The second step is the calculation of the quasiparticle spectral weight. Not all of the partial processes given by Eq. (3.1) with nonzero matrix elements $\gamma(pq)$ in Eq. (3.3) give the nonzero contribution to the spectral density. For example if both the initial $|q\rangle$ and the final states $|p\rangle$ are empty, the spectral weight of the quasiparticle created by the X^{pq} operator is equal to zero. To calculate the spectral weight it is necessary to find the corresponding Green's function.

Now we discuss the quasiparticle statistics. The matrix element $\gamma(pq)$ in Eq. (3.3) is nonzero only when the number of electrons n_p in the final state $|p\rangle$ differs from the number of electrons n_q in the initial state $|q\rangle$ by one:

$$n_p - n_q = 1.$$

According to the properties of the X-operators given in Chapter 1, the corresponding X^{pq} operator is of Fermi type,

$$X_f \cdot X_g = -X_g \cdot X_f \quad \text{when } f \neq g.$$

At any one site, the anticommutator

$$\left\{X_f^{pq}, X_f^{\overset{+}{pq}}\right\}_+ = \left\{X_f^{pq}, X_f^{qp}\right\}_+ = X_f^{pp} + X_f^{qq} \equiv \hat{F}(p,q) \qquad (3.4)$$

is not equal to 1 but is a diagonal operator. A similar situation is known for the spin operators. Thus, the quasiparticle described by the X^{pq} operator has statistics that is more complicated than that of a free electron. It is clear from the definition that the quasiparticle has the electric charge e and spin $S = 1/2$, provided the matrix elements $\langle p|a_{f\lambda\sigma}|q\rangle$. The change of statistics in our approach is connected with the spectral weight redistribution as has been discussed in detail in Chapter 1 for the Hubbard model. For an arbitrary SEC model the situation is the same: the spectral weight of a given quasiparticle is determined by the filling factor $F(p,q) = \langle \hat{F}(p,q) \rangle$ given by Eq. (3.4) and the filling factor is calculated self-consistently via the equation for the chemical potential. For a particular case of the Hubbard model in the half-filled band in the paramagnetic phase at $T = 0$, the filling factors for both the LHB and the UHB are $F(\Omega_+) = F(\Omega_-) = 1/2$. This results in the total number of states for each Hubbard subband being equal to 1, instead of 2 as with the free electrons. Nevertheless, we emphasize that quasiparticles in the Hubbard subbands have both electric charge and spin 1/2. Contrary to the spinless fermion in the external magnetic field, the Hubbard subbands have Zeeman splitting.

Bose type quasiparticles are determined in the similar way. For the Bose excitations the matrix elements in Eq. (1.37) are nonzero only when $n_p - n_q = 0, 2, 4 \ldots$. The Bose type operators X^{pq} commute at different sites and at one site their commutator is given by

$$\left[X_f^{pq}, X_f^{qp}\right]_- = X_f^{pp} - X_f^{qq} \equiv \hat{B}(p,q). \qquad (3.5)$$

The excitons and magnons are the examples of the Bose type excitations in multielectron systems.

3.2 The Exact Intracell Local Green's Function

Due to exact diagonalization, the intracell part of the total Hamiltonian is given by

$$\hat{H}_0 - \mu \hat{N} = \sum_f \sum_p (E_p - n_p \mu) X_f^{pp}, \qquad (3.6)$$

where f is a lattice site, E_p is the energy of the multielectron state $|p\rangle$ where n_p is the number of electrons and μ is a chemical potential. We take the usual definition of the single particle Green's function for electrons or bosons

$$G_{\lambda\lambda',\sigma}(f,t;g,t') = \langle\langle a_{f\lambda\sigma}(t)|a^+_{g\lambda'\sigma}(t')\rangle\rangle. \tag{3.7}$$

With the help of Eq. (3.3) one obtains the X-representation for the Green's function

$$G_{\lambda\lambda',\sigma}(f,t;g,t') = \sum_{\alpha,\beta} \gamma_{\lambda\sigma}(\alpha)\gamma^*_{\lambda'\sigma}(\beta)D_{\alpha\beta}(f,g),$$
$$D_{\alpha\beta}(f,g) = \langle\langle X^\alpha_f | X^{-\beta}_g \rangle\rangle. \tag{3.8}$$

Here we have used the root vectors $\boldsymbol{\alpha}(p,q)$ to denote the pair of initial state $|q\rangle$ and final state $|p\rangle$ (see Chapter 1).

For the diagonal intracell Hamiltonian (3.6), the equation of motion is solved exactly

$$i\dot{X}^\alpha_f = \left[\dot{X}^\alpha_f, \hat{H}_0 - \mu\hat{N}\right]_- = \sum_{m,g}\left[X^{p,q}_f, (E_m - n_m\mu)X^{m,m}_g\right]_-$$
$$= (E_q - E_p - (n_q - n_p)\mu)X^{p,q}_f \equiv \Omega(\alpha)X^\alpha_f, \tag{3.9}$$

where $\Omega(\alpha)$ is the quasiparticle energy (3.1). For the Fermi type quasiparticles $n_q = n_p + 1$, this means that the quasiparticle has the electric charge e. The Bose type quasiparticles may be neutral ($n_p = n_q$) or charged ($n_q - n_p = 2, 4, \ldots$). The Fourier transform of the Green's function (3.7) is equal to

$$D^{(0)}_{\alpha\beta}(f,g) = \langle\langle X^\alpha_f|X^{-\beta}_g\rangle\rangle^{(0)}_E = \frac{\left\langle \left[X^\alpha_f, X^{-\beta}_g\right]_\pm \right\rangle}{E - \Omega(\alpha) + i\varepsilon}\delta_{\alpha\beta}. \tag{3.10}$$

The important difference of the quasiparticle Green's function (3.10) from the free particle Green's function is that the filling factor in the numerator equal, for Fermi particles, to

$$\left\langle \left[X^\alpha_f, X^{-\beta}_g\right]_+ \right\rangle = \delta_{fg}\delta_{\alpha\beta}F(\alpha), \quad F(\alpha) = \left\langle X^{pp}_f \right\rangle + \left\langle X^{qq}_f \right\rangle, \tag{3.11}$$

and, for Bose particles, to

$$\left\langle \left[X^\alpha_f, X^{-\beta}_g\right]_- \right\rangle = \delta_{fg}\delta_{\alpha\beta}B(\alpha), \quad B(\alpha) = \left\langle X^{pp}_f \right\rangle - \langle X^{qq}_f\rangle. \tag{3.12}$$

Thus, in the local representation there are Bose- and Fermi-type quasiparticles with the energy $\Omega(\alpha)$ and the spectral weight $B(\alpha)$ or $F(\alpha)$, correspondingly. The spectral weight is not equal to 1 for all root vectors and depends on temperature, magnetic field and electron concentration. This dependence is the essential feature of quasiparticles in the SEC systems.

The exact local electron Green's function is given by

$$G^{(0)}_{\lambda\lambda',\sigma}(f,g) = \delta_{fg}\delta_{\lambda\lambda'} \sum_\alpha \frac{|\gamma_{\lambda\sigma}(\alpha)|^2 F(\alpha)}{(E - \Omega_\alpha + i\varepsilon)}. \qquad (3.13)$$

Using the exact representation of the single electron operator (3.3), the following sum rule can be formulated

$$\left\langle \left[a_{f\lambda\sigma}, a^+_{f\lambda\sigma}\right]_+\right\rangle = 1 = \sum_\alpha |\gamma_{\lambda\sigma}(\alpha)|^2 F(\alpha). \qquad (3.14)$$

This rule results in the conservation of the total number of states in each single electron band λ:

$$\sum_\sigma \int dE \left(-\frac{1}{\pi}\right) Im G^{(0)}_{\lambda,\sigma}(E + i\varepsilon) = 2, \qquad (3.15)$$

irrespective of the representation for the Green's function.

3.3 The Generalized Tight-Binding Method for Quasiparticle Band Structure Calculations

The operator equivalence of all models considered in Chapter 2 to the multilevel Hubbard model allows one to study all models on equal footing by methods developed for the Hubbard model. The diagram technique will be given below in Chapter 5. Here we restrict ourselves to the simplest approximation of the Hubbard 1 type. It should be mentioned that in the cluster perturbation theory this approximation means neglecting correlations between different cells (clusters) but it concerns not all interatomic correlations. Some part of the interatomic correlations that is included into the cell is treated exactly on the level of the cell Hamiltonian H_0. Therefore, the accuracy of the Hubbard 1 approximation in the cluster perturbation theory is better than in the usual decoupling procedure in the atomic perturbation theory. At least quasiparticle band structure of the undoped CuO_2 layer (Chapter 8) calculated by this method is in good agreement with the results of the exact diagonalization and QMC study of $t-t'-t''-J$ model.

The better accuracy of the cluster perturbation theory versus the atomic perturbation theory results from the matrix elements of the intercluster perturbations dependence on the number of atoms N_a in the cluster (Ovchinnikov and Petrakovsky 1991). A cluster eigenvector contains the normalization factor, $|p\rangle \sim N_a^{-1/2}$. Single particle operators a, a^+ according to Eq. (3.3) contain the factor N_a^{-1}. The kinetic energy (intercluster hopping) term $t_{ij} a_i^+ a_j$ renormalized as $t_{ij} X_i X_j / N_a^2$. The intercluster Coulomb matrix elements $V a_1^+ a_2^+ a_3 a_4$ contain the factor N_a^{-4}. For the typical value $N_a = 5\text{--}10$, it is evident that the most essential intercluster perturbation is given by the hopping. For the exact solution of the multiband Hubbard model these renormalizations are not important because the number of different cluster energy levels and the scattering channels are increasing equally. But for the approximate low energy solution the fast decreasing of the intercluster Coulomb interaction is important. Thus the main perturbation we consider here is the kinetic energy term

$$H_1 = \sum_{fR} \sum_{\lambda\lambda'\sigma} t_{\lambda\lambda'}(R) a_{f\lambda\sigma}^+ a_{f+R,\lambda'\sigma} + \text{h.c.} \qquad (3.16)$$

Further simplification of notations is done below. Due to the finite number of root vectors one can numerate them: $\alpha_1, \alpha_2, \ldots, \alpha_m$. Here index m is just the number of quasiparticles. Therefore, instead of the initial notation for the X-operator we obtain the following

$$a_{f\lambda\sigma} = \sum_m \gamma_{\lambda\sigma}(m) X_f^m, \quad a_{f\lambda\sigma}^+ = \sum_m \gamma_{\lambda\sigma}^*(m) X_f^{+m} \qquad (3.17)$$

and the perturbation Hamiltonian H_1 is given by

$$H_1 = \sum_{f,R} \sum_{\lambda,\lambda',\sigma} \sum_{m,n} t_{\lambda\lambda'}(R) \gamma_{\lambda\sigma}^*(m) \gamma_{\lambda'\sigma}(n) X_f^{+m} X_{f+R}^n + \text{h.c.} \qquad (3.18)$$

Of course, there is a table of correspondence $m \leftrightarrow \alpha_m \leftrightarrow (p,q)$. We define the Fourier transform of the X-operators as follows

$$X_f^m = \frac{1}{N} \sum_k X_k^m \exp(ikR_f), \quad X_k^m = \sum_f X_f^m \exp(-ikR_f), \qquad (3.19)$$

and we write the perturbation H_1 in reciprocal space

$$H_1 = \sum_{k,m,n} \sum_{\lambda,\lambda',\sigma} t_{\lambda\lambda'}(k) \gamma_{\lambda\sigma}^*(m) \gamma_{\lambda'\sigma}(n) X_k^{+m} X_k^n + \text{h.c.} \qquad (3.20)$$

Here, $t_{\lambda\lambda'}(k)$ is the Fourier transform of the hopping integral usual for the tight-binding approach

$$t_{\lambda\lambda'}(k) = \sum_R t_{\lambda\lambda'}(R)\exp(ikR). \qquad (3.21)$$

It is convenient also to write down the Hamiltonian H_1 and the hopping integrals in the quasiparticle representation

$$t_{mn}(k) = \sum_{\lambda,\lambda',\sigma} t_{\lambda\lambda'}(k)\gamma^*_{\lambda\sigma}(m)\gamma_{\lambda'\sigma}(n),$$

$$t_{mn}(R) = \sum_{\lambda,\lambda',\sigma} t_{\lambda\lambda'}(R)\gamma^*_{\lambda\sigma}(m)\gamma_{\lambda'\sigma}(n), \qquad (3.22)$$

$$H_1 = \sum_{f,R}\sum_{m,n} t_{m,n}(R) X_f^{+m} X_{f+R}^n = \sum_k \sum_{m,n} t_{m,n}(k) X_k^{+m} X_k^n + \text{h.c.} \qquad (3.23)$$

Equation (3.23) is the same for the single-band Hubbard model and the multiband model, the only difference is the number of possible values of the indexes m and n.

The exact equation of motion for the operator X_f^m is given by

$$i\dot{X}_f^m = [X_f^m, H_0 + H_1]_- = \Omega_m X_f^m + [X_f^m, H_1]_-. \qquad (3.24)$$

Evaluating the commutator in Eq. (3.24) results in

$$[X_f^m, H_1] = 2 \sum_{R,n_1,n_2} t_{n_1 n_2}(R) \left(E_f^{mn_1} X_{f+R}^{n_2} + D_f^{mn_2} X_{f+R}^{+n_1} \right) \qquad (3.25)$$

with operators E_f and D_f describing neutral and doubly charged local bosons, correspondingly

$$\left\{ X_f^m, X_g^{+n} \right\}_+ = \delta_{fg} F_f^{m,n}, \quad \{X_f^m, X_g^n\}_+ = \delta_{fg} D_f^{m,n}. \qquad (3.26)$$

For example, in the t–J model with only one Fermi type quasiparticle we have

$$m = 0, \quad \alpha_0 = (0,\sigma) \quad \text{and} \quad X_f^m = X_f^{0\sigma}.$$

Besides the diagonal Bose operators, the spin fluctuation operator appears

$$E_f^{m,n} = \delta_{\sigma\sigma'} X_f^{00} + X_f^{\sigma',\sigma}.$$

For the Hubbard model with $\alpha_0 = (0, \sigma)$ and $\alpha_1 = (-\sigma, 2)$, the Bose excitations in the right part of Eq. (3.25) are given by 2×2 matrices

$$E_f^{m,n} = \begin{pmatrix} \delta_{\sigma\sigma'} X_f^{00} + X_f^{\sigma'\sigma} & 0 \\ 0 & \delta_{\sigma\sigma'} X_f^{22} + X_f^{-\sigma,-\sigma'} \end{pmatrix},$$

$$D_f^{m,n} = \begin{pmatrix} 0 & \delta_{\sigma,-\sigma'} X_f^{02} \\ \delta_{\sigma,-\sigma'} X_f^{02} & 0 \end{pmatrix}.$$

Here, the doubly charged excitations are additional excitations to those appearing in the usual t–J model. For a concrete model the explicit form of the neutral and charged bosons is obtained straightforwardly using the definition (3.26).

The Hubbard 1 approximation means that the local Bose type operators $\hat{E}_f$ and $\hat{D}_f$ are substituted with their average values. For the space uniform normal state

$$E_f^{m,n} \to \left\langle E_f^{m,n} \right\rangle = \delta_{mn} F(m), \quad D_f^{m,n} \to \left\langle D_f^{m,n} \right\rangle = 0,$$

the filling factor is

$$F_f(m) \equiv \langle F_f(\alpha_m(p,q)) \rangle = \left\langle X_f^{p,p} \right\rangle + \left\langle X_f^{q,q} \right\rangle.$$

For the X-operator Green's function

$$D^{m,n}(f,g) \equiv \left\langle\!\left\langle X_f^m \middle| X^n \right\rangle\!\right\rangle_E, \qquad (3.27)$$

the Hubbard 1 decoupling results in the following integral equation

$$(E - \Omega_m) \left\langle\!\left\langle X_f^m \middle| X^{m'} \right\rangle\!\right\rangle$$
$$= F_f(m) \delta_{fg} \delta_{mm'} + \sum_{n,R} t_{mn}(R) F_f(m) \left\langle\!\left\langle X_{f+R}^n + X_{f-R}^n \middle| X_g^{m'} \right\rangle\!\right\rangle_E.$$
$$(3.28)$$

For the space uniform system, the Fourier transform of the function (3.27) is determined by

$$D^{mm'}(k) = D_0^m \delta_{mm'} + D_0^m \sum_n t_{mn}(k) D^{nm'}(k), \qquad (3.29)$$

where $D_0^m = F(m)/(E - \Omega_m)$ is the local intracell Green's function (3.10). In matrix notation, Eq. (3.29) has the following form

$$\hat{D}(k) = \hat{D}_0 + \hat{D}_0 \hat{t}(k)\hat{D}(k), \tag{3.30}$$

and its solution

$$\hat{D}^{-1}(k) = \hat{D}_0^{-1} - \hat{t}(k) \tag{3.31}$$

has a typical mean field form. In the diagram technique for the X-operator, it was shown explicitly that the Hartree–Fock solution for the Green's function (3.27) is equal to the results of the Hubbard 1 approximation (Zaitsev 1976; Vedyaev and Ivanov 1981, 1982). The components of the Green's function (3.27) are given by the inverse matrix components

$$D^{m,n}(k, E) = \left(\frac{\delta_{mn}(E - \Omega_m)}{F_{(m)} - t_{mn}(k)}\right)^{-1}. \tag{3.32}$$

The dispersion equation for the Fermi type quasiparticles has the following form

$$\det \|\delta_{mn}(E - \Omega_m)/F(m) - t_{mn}(k)\| = 0. \tag{3.33}$$

This equation is similar to the conventional tight-binding dispersion equation but has three important differences: firstly, here the indexes m, n do not mean the free electron bands but the quasiparticle bands. Secondly, the local single particle energy Ω_m is not the free electron one but is determined as the difference of two multielectron terms (3.1) with $(N+1)$ and N-electrons. Thirdly, the effective hopping integral is a product of the single electron hopping $t(k)$ and the filling factor $F(m)$, resulting in the concentration, the temperature as well as the field dependence of the quasiparticle band structure. Therefore, we call this method the generalized tight binding (GTB) method.

The single electron Green's function (3.7) is equal to

$$G_{\lambda,\lambda',\sigma}(k, E) = \sum_{m,n} \gamma_{\lambda\sigma}(m)\gamma^*_{\lambda'\sigma}(n)D^{m,n}(k, E). \tag{3.34}$$

The single particle spectral density function is given by

$$A_\sigma(k, E) = -\frac{1}{\pi}\sum_{\lambda,m,n} \gamma_{\lambda\sigma}(m)\gamma^*_{\lambda\sigma}(n)JmD^{m,n}(k, E + i\delta). \tag{3.35}$$

A single particle density of states is also of interest

$$N_\sigma(E) = \frac{1}{N} \sum_k A_\sigma(k, E). \tag{3.36}$$

For the two-sublattice structure in the antiferromagnetic phase, the same approach may be applied straightforwardly (Ovchinnikov 1987). For the two sublattices A and B with sites $f \in A, g \in B$ the local excitation energies and filling factor depend on the sublattice index, $\Omega_{A,B}(m)$ and $F_{A,B}(m)$. For the matrix Green's function

$$\hat{D} = \begin{pmatrix} D_{A,A}, & D_{A,B} \\ D_{B,A}, & D_{B,B} \end{pmatrix},$$

the Hubbard 1 solution

$$\hat{D}_{(k)}^{-1} = \hat{D}_0^{-1} - \hat{t}(k), \tag{3.37}$$

has the same form as Eq. (3.31), but both bare Green's function $\hat{D}_0$ and hopping term $\hat{t}$ are now matrices in the index m, n and sublattice indexes A, B:

$$\hat{D}_0^{-1} = \begin{pmatrix} \frac{E-\Omega_A(m)}{F_A(m)}, & 0 \\ 0, & \frac{E-\Omega_B(m)}{F_B(m)} \end{pmatrix}, \tag{3.38}$$

$$\hat{t}(k) = \begin{pmatrix} \sum_{R_1} t_{mn}^{AA}(R_1) \exp(ikR_1), & \sum_{R_2} t_{mn}^{AB}(R_2) \exp(ikR_2) \\ \sum_{R_2} t_{mn}^{BA}(R_2) \exp(ikR_2), & \sum_{R_1} t_{mn}^{BB}(R_1) \exp(ikR_1) \end{pmatrix}. \tag{3.39}$$

In the last equation we have denoted the vector of the intrasublattice hopping by R_1 and the vector of the intersublattice hopping by R_2. The GTB calculations for antiferromagnetic semiconductors in the framework of the s–d exchange model have been done by Ovchinnikov (1987). The same calculations for copper oxides in the framework of the multiband p–d model have been carried out by Ovchinnikov (1995) in the undoped case and by Gavrichkov et al (2000) for the hole doped systems. The quasiparticle band structure of copper oxides will be considered later in Chapter 8.

3.4 The Concentration Dependence of the Quasiparticle Band Structure

The presence of the filling factor in the GTB dispersion equation (3.33) results in concentration dependent band structure. It is worth mentioning that at the conventional single electron approach a rigid band behavior usually takes place, i.e. the increasing (decreasing) electron concentration results in the corresponding shift of Fermi energy. To obtain an impurity level in the forbidden energy gap of a semiconductor some disorder like the fluctuations of the crystal potential or the interatomic hopping integrals is required. Here we show that in the insulator with the Mott–Hubbard or charge-transfer gap a new specific feature for the SEC system mechanism of the impurity-like band formation appears which does not require any disorder in the system (Ovchinnikov 1992). The physical origin of the appearance of the new impurity-like state with the introduction of doping is the redistribution of spectral weight and filling factors. Here we consider this mechanism in detail.

In the undoped insulator material like La_2CuO_4 the number of electrons (holes) per cell is an integral value. To add and to remove one electron, the $n_e + 1$ and $n_e - 1$ electron configurations are to be considered. For simplicity we can assume that $n_e = 1$ in the undoped case. The energy levels of relevant states are shown in the Fig. 3.1(b). At zero temperature only the ground $n_e = 1$ term $|1, 0\rangle$ with energy $E_0(1)$ is occupied, all the other are empty. Only quasiparticles having the $|1, 0\rangle$ term either as the initial or final states have nonzero filling factors. If one were to take into account only one excited term in the $n_e = 1$ ($E_1(1)$) and $n_e = 2$ ($E_1(2)$) sectors of the Hilbert space then there are 3 Fermi type quasiparticles with nonzero filling factors:

$$\Omega_0 = E_0(1) - E_0(0),$$
$$\Omega_1 = E_0(2) - E_0(1), \quad (3.40)$$
$$\Omega_2 = E_1(2) - E_0(1).$$

The density of states in the zeroth order in hopping is equal to

$$N_0(E) = \sum_{i=0}^{2} \delta(E - \Omega_i).$$

The intersite hopping results in the formation of the bands, Ω_0 forms the LHB and Ω_1 forms the UHB. Let us assume that

$$\max\{t(k)\} \ll U \sim \Omega_1 - \Omega_0$$

and the system is in the insulator state with the top of the valence band at the energy E_b and the bottom of the empty conductivity band at the energy E_c, separated by the insulator gap E_g:

$$E_c - E_b = E_g.$$

To calculate the position of Fermi level we have to consider the spin structure of $(n_e - 1), n_e, n_e + 1$ multiplets. For $n_e = 1$ each term is a spin doublet while among the $n_e = 2$ terms there are singlets and triplets. When $n_e = 1$, the occupation number $n_{1,0,\sigma} = 1/2$ and for all the other terms the occupation numbers are equal to zero. So, the nonzero filling factors are

$$F(\alpha_0) = F(\alpha_1) = F(\alpha_2) = 1/2.$$

The LHB with respect to for spin contains 1 electron and is filled while the UHB is empty, the Fermi level lying inside the gap.

Due to doping, the electron concentration becomes equal to $n_e = 1 + x$. From the equation for the chemical potential

$$n_e = \sum_\gamma \left\langle X_i^{1\gamma,1\gamma} \right\rangle + 2 \sum_\gamma \left\langle X_i^{2\gamma,2\gamma} \right\rangle, \tag{3.41}$$

it is clear that at $T = 0$ two ground state terms are partially filled: $E_0(1)$ with probability $\left\langle X_i^{1,0;1,0} \right\rangle = 1 - x$, and the two electron ground state $E_0(2)$ with the probability $\left\langle X_i^{2,0;2,0} \right\rangle = x$. A nonzero filling factor appears now for a new quasiparticle $\Omega_3 = E_0(2) - E_1(1)$. This filling factor is proportional to x. Our claim concerning new quasiparticle band with spectral density proportional to the doping concentration x has a general character but to calculate the number of states in the doped system one is required to know the matrix elements for the quasiparticles involved, so we consider the two orbital Hubbard model with the local part of the Hamiltonian given by

$$H_0(i) = \sum_\sigma (\varepsilon_1 n_{1\sigma} + \varepsilon_2 n_{2\sigma}) + U_1 n_{1\uparrow} n_{1\downarrow} + U_2 n_{2\uparrow} n_{2\downarrow}$$

$$+ V n_{1\sigma} n_{2\sigma'} - J S_1 S_2. \tag{3.42}$$

The local eigenstates of the Hamiltonian (3.42) are the following:

(a) $n = 0$ vacuum state $|0\rangle$,
(b) $n = 1$ two doublets $|\lambda\sigma\rangle$, where $\lambda = 1, 2$ is the orbital index,
(c) $n = 2$ three singlets and one triplet

$$|S, \lambda\rangle = \alpha^+_{\lambda\uparrow}\alpha^+_{\lambda\downarrow}|0\rangle; \quad |S, 0\rangle = 2^{-1/2}(\alpha^+_{1\uparrow}\alpha^+_{2\downarrow} - \alpha^+_{1\downarrow}\alpha^+_{2\uparrow})|0\rangle,$$
$$|T, 2\sigma\rangle = \alpha^+_{1\sigma}\alpha^+_{2\sigma}|0\rangle, \quad |T, 0\rangle = 2^{-1/2}(\alpha^+_{1\uparrow}\alpha^+_{2\downarrow} + \alpha^+_{1\downarrow}\alpha^+_{2\uparrow})|0\rangle, \quad (3.43)$$

(d) $n = 3$ two doublets $|3, \lambda\sigma\rangle$

$$|3, 1\sigma\rangle = \alpha^+_{1\sigma}\alpha^+_{2\uparrow}\alpha^+_{2\downarrow}|0\rangle, \quad |3, 2\sigma\rangle = \alpha^+_{1\uparrow}\alpha^+_{1\downarrow}\alpha^+_{2\sigma}|0\rangle.$$

All possible quasiparticles with nonzero matrix elements $\langle p|a_{\lambda\sigma}|q\rangle$ are presented in the Table 3.1. The last two columns of Table 3.1 contain the filling factors F_S and F_T for two possible $E_0(2)$ spin states, the singlet and triplet states correspondingly. It is seen from Table 3.1 that a nonzero spectral weight appears with doping for the quasiparticles with numbers 3, 10 and 14 in the case of the singlet two electron ground state and with numbers 6, 7, 11, 12, 15, 16 for the triplet $n_e = 2$ ground state. It should be noted that with doping the Fermi type quasiparticles appear to involve three-electron terms which are not shown in the Fig. 3.1(b):

$$\Omega_4 = E_0(3) - E_0(2), \quad \Omega_5 = E_1(3) - E_0(2).$$

With the help of Table 3.1 it is easy to see that the sum rule

$$\sum_\sigma \sum_\alpha \left(|\gamma_1(\alpha)|^2 + |\gamma_2(\alpha)|^2\right) F(\alpha) = 4$$

takes place for all x.

For the quasiparticle Ω_1 involving the excited $n_e = 1$ term

$$\Delta E = \Omega_1 - \Omega_3 = E_1(1) - E_0(1) > 0,$$

the level Ω_3 lies below the Ω_1 and is inside the gap E_g under the following conditions

$$E_b < \Omega_3 < E_c. \quad (3.44)$$

The number of states for the quasiparticle Ω_3 is given by $F(\alpha_3) \sim x$. It means that in regular system with the non-integral number of electrons the impurity-like level appears inside the gap. The dispersion of this level is given by the same filling factor $F(\alpha_3) \sim x$. It is small at $x \ll 1$. It is also

Table 3.1 Different quasiparticles, root vectors, matrix elements, and filling factors of singlet F_S and triplet F_T double-electron states for the two-band Hubbard model

Quasiparticle	Root vector	$\gamma_1(\alpha)$	$\gamma_2(\alpha)$	$F_S(\alpha)$	$F_T(\alpha)$
0	$(0; 1\sigma)$	1	0	$(1-x)/2$	$(1-x)/2$
1	$(0; 2\sigma)$	0	1	0	0
2	$(1-\sigma; S, 0)$	0	$-\sigma\sqrt{2}$	$(1+x)/2$	$(1-x)/2$
3	$(2-\sigma; S, 0)$	$\sigma\sqrt{2}$	0	x	0
4	$(1, \sigma; T, 2\sigma)$	0	1	$(1-x)/2$	$1/2 - x/6$
5	$(1, -\sigma; T, 0)$	0	$-1/\sqrt{2}$	$(1-x)/2$	$1/2 - x/6$
6	$(2, \sigma; T, 2\sigma)$	1	0	0	$x/3$
7	$(2, -\sigma; T, 0)$	$1/\sqrt{2}$	0	0	$x/3$
8	$(1, -\sigma; S, 1)$	2σ	0	$(1-x)/2$	$(1-x)/2$
9	$(2, -\sigma; S, 2)$	0	2σ	0	0
10	$(S, 0; 3, 1, \sigma)$	0	$-1/\sqrt{2}$	x	0
11	$(T, -2\sigma; 3, 1, -\sigma)$	0	-2σ	0	$x/3$
12	$(T, 0; 3, 1, \sigma)$	0	$-\sigma\sqrt{2}$	0	$x/3$
13	$(S, 2; 3, 1, \sigma)$	1	0	0	0
14	$(S, 0; 3, 2, \sigma)$	$1/\sqrt{2}$	0	x	0
15	$(T, -2\sigma; 3, 2, -\sigma)$	2σ	0	0	$x/3$
16	$(T, 0; 3, 2, \sigma)$	$\sigma\sqrt{2}$	0	0	$x/3$
17	$(S, 1; 3, 2, \sigma)$	0	1	0	0

true for levels Ω_4 and Ω_5. If the condition (3.44) does not hold for some of the levels Ω_i ($i = 3, 4, 5$), the corresponding level lies in the band and becomes virtual.

To find the Fermi level we restrict ourselves to the case of the singlet $n_e = 2$ state $|2, 0\rangle$. With regard to spin, the number of states in the LHB Ω_0 is equal to $1 - x$, while for the impurity like level Ω_3 the number of states is equal to $2x$. So the total number of electrons $n_e = 1 + x$ is equal to the number of states both in the LHB and the impurity level Ω_3, which are filled. The UHB is empty, and the Fermi level is inside the gap. An increasing doping requires the hopping to be included and as a result the impurity band will appear. The evolution of the impurity level in the band will be considered for the multiband p–d model of cuprates in Chapter 8.

A new mechanism of the impurity level formation in SEC systems is caused by the contribution of the higher energy excited states to the single electron spectral density. Such contribution is absent in the undoped case. Of course in experiments with doped compounds the ion substitution not only changes the electron concentration, some impurity fluctuations appear (the diagonal or the nondiagonal disorder), and both mechanisms contribute to the impurity level formation.

3.5 Ab Initio Approach to the Quasiparticle Band Structure of Strongly Correlated Electron Systems

The above given GTB method has two important shortcomings. The main one is a model-dependent approach of the GTB method with a model Hamiltonian with a set of single-electron and two-electron matrix elements. These matrix elements have to be calculated separately or obtained by fitting some experimental data. The more realistic the model, the more is the number of input parameters in the model Hamiltonian. A second shortcoming is that of using the Hubbard 1 approximation that may be valid only in the atomic limit with a bandwidth $W \ll U$. For many 3d- and 4f-metal oxides that are Mott–Hubbard or charge-transfer insulators this approximation seems to be quite reasonable.

To overcome the first shortcomings an *ab initio* approach should be used. A conventional band theory is based on the density functional theory (DFT) by Hohenberg and Kohn (1964) and on the Local Density Approximation (LDA) to DFT (Kohn and Sham 1965). In spite of great success of the LDA for metallic systems it appears to be inadequate for strongly correlated systems. A well-known example is the discrepancy of the metallic LDA band structure and the insulator ground state of La_2CuO_4. Several approaches to incorporate SEC effects in the LDA method are known, for example LDA + U (Anisimov *et al* 1991) and LDA-SIC (Svane and Gunnarsson 1990). Both methods result in the correct antiferromagnetic insulator ground states for La_2CuO_4 contrary to LDA, but the insulating gap is formed by the local single-electron states splitting over spin or orbital polarization. In such approach a paramagnetic half-filled Hubbard model will be a metal even in the strong correlation regime $U \gg W$. The main effect of SEC is the spectral weight distribution that results in the Mott–Hubbard gap in the paramagnetic phase.[1]

This effect is incorporated in the LDA++ approach (Lichtenstein and Katsnelson 1998; Katsnelson and Lichtenstein 2000) where a syntesis of the LDA and the dynamical mean field theory, described above in Section 1.7, is given. The abbreviation LDA++ means LDA + U + Σ, that is the LDA *ab initio* calculations with electron correlations (U) with the help of the frequency dependent self-energy $\Sigma(\omega)$ in the framework of the dynamical mean field theory. In the weak correlation limit $W \gg U$, the LDA++

---

[1] One of the authors (S. G. Ovchinnikov) thanks A. Georges for the fruitful discussion of this problem.

is equivalent to the conventional diagram perturbation theory for Fermi particles. In the opposite limit of strong correlations $W \ll U$, the LDA++ use the Hubbard 1 approximation. In the intermediate regime $U \sim W$, the self-energy is calculated by the dynamical mean field approach. Recent paper by Chitra and Kotliar (2000) provides a theoretical underpinning to the LDA++ method. In general, the LDA++ is a very important step in the problem of *ab initio* band theory of SEC systems. Nevertheless there are some unsolved problems in the LDA++. One of them is the two-electron matrix elements, a problem of "effective U", that are not calculated *ab initio* and are the fitting parameters. The second problem is a double-counting of the Coulomb interaction. To avoid this problem the effective potential is subtracted, this procedure can be justified for metals but not for Mott–Hubbard insulators.[2]

The other approach to *ab initio* band structure calculation in SEC systems is developed by the Uppsala University group (Sandalov *et al* 2000; Lundin *et al* 2000). They use the X-operator perturbation method (Sandalov *et al* 2000) to calculate the total energy of localized electrons (for example, f-electrons in rare-earth metals) that in the Sham (1985) theory is given by a sum of skeleton diagrams. The exchange–correlation potential $v_{xc}[\rho(r)]$ has the same analytical expression over the charge density $\rho(r)$ as in the LDA with the difference that $\rho(r)$ is calculated for SEC system in X-operator representation. The electron Green's function is obtained using Kadanoff and Baym (1962) idea in a formal closed expression with the self-energy given by a variational derivative. For practical calculation the X-operator diagram technique (see Chapter 5 below) is used. This approach also has a problem of "effective U" that is considered as a fitting parameter.

In the following we will discuss a possible combination of our GTB method with *ab initio* approach. We start with the usual Hamiltonian of valence electrons in a second quantization from

$$H_\nu = \sum_\sigma \int dr \psi_\sigma^+(r) \left(\frac{p^2}{2m} - \sum_i \frac{(Z_i - n_i^{(c)})e^2}{|r - R_i|} \right) \psi_\sigma(r)$$
$$+ \frac{1}{2} \sum_{\sigma_1 \sigma_2} \int dr_1 dr_2 \psi_{\sigma_1}^+(r_1) \psi_{\sigma_2}^+(r_2) \frac{e^2}{r_{12}} \psi_{\sigma_2}(r_2) \psi_{\sigma_1}(r_1) \equiv H_0 + H_{\text{int}}.$$
(3.45)

---

[2]S. G. Ovchinnikov is thankful to M. Katsnelson for very useful discussion of the LDA++ method.

Let us suppose that by the conventional LDA method we find a Bloch wavefunction $\Psi_{k\lambda\sigma}(r)$ for an electron with a wavevector k, band index λ, and spin projection σ. Then we determine a Wannier function as usual by the Fourier transform of the Bloch function

$$\Psi_{f\lambda\sigma}(r) = \frac{1}{N}\sum_k e^{-ikR_f}\Psi_{k\lambda\sigma}(r). \qquad (3.46)$$

and write down the Hamiltonian (3.45) in the Wannier representation.

$$\begin{aligned}H_\nu &= \sum_{f\lambda\sigma}\varepsilon_{f\lambda}a^+_{f\lambda\sigma}a_{f\lambda\sigma} + \sum_{\substack{f_1 f_2\\ \lambda_1\lambda_2\sigma}}t^{\lambda_1\lambda_2}_{f_1 f_2}a^+_{f_1\lambda_1\sigma}a_{f_2\lambda_2\sigma}\\ &+ \sum_{\substack{f_1 f_2 f_3 f_4\\ \lambda_1\lambda_2\lambda_3\lambda_4\\ \sigma,\sigma'}}\left\langle f_1\lambda_1, f_2\lambda_2\left|\frac{e^2}{r_{12}}\right|f_4\lambda_4, f_3\lambda_3\right\rangle a^+_{f_1\lambda_1\sigma}a^+_{f_2\lambda_2\sigma'}a_{f_3\lambda_3\sigma'}a_{f_4\lambda_4\sigma}.\end{aligned}$$

(3.47)

All matrix elements in Eq. (3.47) can be calculated, at least in principle. We do not consider here practical problems arising in calculations with the Wannier function[3] (Zak 1981). So we neglect three-site and four-site Coulomb matrix elements in the Eq. (3.47) and believe that we can calculate all two-site matrix elements. Then we can use a cluster perturbation theory, where the Hamiltonian is written as a sum of the intracluster (cell) part H_c and the intercluster (cell–cell) part H_{cc},

$$H_c = \sum_f H_f, \quad H_{cc} = \sum_{fg}H_{fg}. \qquad (3.48)$$

The multielectron molecular orbitals (MEMO) of the isolated cell with the Hamiltonian H_f are found by the configuration interaction (CI) method separately in each sector of the Hilbert space with definite number of electrons n in the cell. For copper oxides, where hole representation is more convenient then electron one, the corresponding sectors and configurations are the following:

(1) Vacuum sector, $n_h = 0$, configurations Φ_{0i}:
   $Cu(4s^0 3d^{10})O(2p^6), Cu(4s^1 3d^9)O(2p^6), Cu(4s^1 3d^{10})O(2p^5)$.

---

[3] Authors are thankful to V.I. Zinenko for the useful remarks and discussion of this problem.

(2) Single-hole sector, $n_h = 1$, configurations Φ_{1i}:
$\text{Cu}(4s^0 3d^9)\text{O}(2p^6), \text{Cu}(4s^0 3d^{10})\text{O}(2p^5), \text{Cu}(4s^1 3d^8)\text{O}(2p^6),$
$\text{Cu}(4s^1 3d^9)\text{O}(2p^5).$
(3) Two-hole sector, $n_h = 2$, configurations Φ_{2i}:
$\text{Cu}(4s^0 3d^9)\text{O}(2p^5), \text{Cu}(4s^0 3d^8)\text{O}(2p^6), \text{Cu}(4s^0 3d^{10})\text{O}(2p^5)\text{O}(2p^5)$, etc.

The corresponding MEMO in the each sector with n electrons can be written as

$$|\Psi_{n,j}\rangle = \sum_i c_{ni}^{(j)} |\Phi_{n,i}\rangle. \quad (3.49)$$

Here an index j denotes a ground and excited states in the given sector n. All MEMO form the orthonormalized basis

$$\langle \Psi_{n_1,j_1} | \Psi_{n_2,j_2} \rangle = \delta_{n_1,n_2} \delta_{j_1,j_2},$$

that allows one to use them in the construction of the X_f-operators in the cell f:

$$X_f^{n_1 j_1, n_2 j_2} \equiv |\Psi_{n_1 j_1}\rangle\langle\Psi_{n_2 j_2}|. \quad (3.50)$$

In the X-representation the intracell part of the Hamiltonian is diagonal

$$H_c = \sum_{fnj} E_{nj} X_f^{nj,nj},$$

and the intercell part H_{cc} has a form of the generalized Hubbard model (2.43). Finally, the quasiparticle band theory can be treated by GTB method (Section 3.3). In this approach all computational problems are shifted to the calculation of single-electron and two-electron matrix elements of the Hamiltonian (3.47) in the Wannier representation.

Summarize the discussion of different *ab initio* approaches we can conclude that the full and complete theory of electronic structure of SEC systems is absent now. Nevertheless some interesting and promising examples of a synthesis of the *ab initio* approach and SEC effects have appeared recently.

3.6 The Generalized Tight-Binding Method and the Exact Lehmann Representation

Several new fascinating ideas have been proposed to treat strongly correlated electron systems (SCES): Luttinger liquid, spinon-holon, slave-boson, slave-fermion, parastatistics etc. Many of them use the idea of spin–charge separation for electrons in SCES that results from the exact solution of the one dimensional l Hubbard model. However the dynamical mean field theory (DMFT) accurate in the large space dimension limit has no indications on spin–charge separation.

Several perturbation approaches on t/U starts from the Hubbard (1965) paper use projection operators or the Hubbard X-operators technique. These approaches include strong Coulomb interactions in the Hamiltonian $H(0)$ and treat the interatomic hopping by a perturbation method. The main conclusion of these perturbation methods is that electron in SCES is considered as a sum of quasiparticles (QP) with charge e, spin 1/2 renormalized energy and spectral weight (Fulde 1991). Application of Hubbard ideas to realistic multiorbital models of transition metal oxides results in a generalized tight-binding (GTB) method.

We use here the exact Lehmann representation for a single electron Green's function (GF) to compare the notion of the quasiparticle in SCES developed in the perturbation approach to the exact one. The Lehmann representation treats an electron as a sum of QP with spin $S = 1/2$, charge e and renormalized energy and spectral weight without any indications on spin–charge separation. We show that the GTB method gives the practical realization of the Lehmann representation. Some results of GTB calculations of the QP energy spectra for copper oxides, its concentration evolution from the undoped insulator to the optimally doped metal, and comparison to ARPES data are given below.

Single electron GF $G_\sigma = \langle\langle a_{k\sigma} | a_{k\sigma}^+ \rangle\rangle_\omega$ can be written as

$$G_\sigma(k,\omega) = \sum_m \left(\frac{A_m(k,\omega)}{\omega - \Omega_m^+} + \frac{B_m(k,\omega)}{\omega - \Omega_m^-} \right), \quad (3.51)$$

where the QP energies are given by

$$\Omega_m^+ = E_m(N+1) - E_0(N) - \mu, \quad \Omega_m^- = E_0(N) - E_m(N-1) - \mu,$$

and the QP spectral weight is equal to

$$A_m(k,\omega) = |\langle 0,N|a_{k\sigma}|m,N+1\rangle|^2, \quad B_m(k,\omega) = |\langle m,N-1|a_{k\sigma}|0,N\rangle|^2.$$

Here $|m,N\rangle$ is the m-th eigenstate of the system with N electrons,

$$H|m,N\rangle = E_m|m,N\rangle.$$

The index m numerates QP with spin $S = 1/2$, electrical charge e, energy $\Omega_m^+(\Omega_m^-)$, and spectral weight $A_m(B_m)$. Equation (3.51) may be considered as a sum over different QP bands with m be a QP band index. In practical calculations the Lehmann representation is useless because the multielectron eigenstates $|m, N\rangle$ for the crystal are not known.

At finite temperature the Lehmann representation for the GF can be written in the following way

$$G_\sigma^R(k, \omega) = \sum_{mn} W_n \frac{A_{mn}(k, \omega)}{\omega - \Omega_{mn}^+ + i0}(1 + e^{-\Omega_{mn}^+/T}). \qquad (3.52)$$

Here, the statistical weight of a state $|n\rangle$:

$$\Omega_{mn}^+ = E_m(N+1) - E_n(N) - \mu,$$

is determined by the Gibbs distribution with the thermodynamical potential Ω:

$$W_n = \exp\{(\Omega - E_n + \mu N)/T\}.$$

At non-zero temperature both the ground state $|0, N\rangle$, and the excited states $|n, N\rangle$ are populated. Thus a quasiparticle is numerated by a pair of indexes m, n. It is a single electron excitation in the multielectron system due to the addition of one electron to an initial N-electron state $|n, N\rangle$ and a final $(N + 1)$-electron state $|m, N + 1\rangle$.

The QP bands of undoped and hole doped CuO_2 layer has been calculated by GTB method in Gavrichkov *et al* (2000) using CuO_6 unit cell CuO_4Cl_2 for $Sr_2CuO_2Cl_2$. In the undoped antiferromagnetic insulator there is a charge transfer gap, the dispersion of the top of the valence band is in good agreement with ARPES data on $Sr_2CuO_2Cl_2$ (Wells *et al* 1995). At the top of the valence band an impurity-like band appears due to spin fluctuations and doping. A pseudogap between the impurity-like band and the main valence band has a dispersion similar to a "remnant Fermi surface" in $Sr_2CuO_2Cl_2$ and to a pseudogap in the underdoped Bi-2212 samples (Gavrichkov *et al* 2001). The Fermi level is pinned inside the impurity-like band and depends very weakly on doping in the underdoped region. QP spectral density, density of states, and the Fermi surface calculated at different hole concentrations show the evolution of the QP band structure from the undoped antiferromagnetic charge-transfer insulator to the optimally doped paramagnetic metal. These results will be discussed in detail in Chapter 8.

In the perturbation approach of GTB the structure of GF is the same as in the exact Lehmann representation. There is no spin-charge splitting of electron in SCES. There is a splitting of electron given by Eq. (3.3) on different QP bands characterized by spectral weigh redistribution over QP bands. Only summing all QP spectral weights will one get free electron spectral weight. Thus spectral weight splitting and removing of its large part far away from the Fermi level is the most essential effect of strong correlations.

Chapter 4

Unitary Transformation Method in Atomic Representation

An effective method of the cell Hamiltonian H_f diagonalization is discussed in this chapter. In the theory of the single-ion anisotropic Heisenberg magnets with small S, unitary transformation has been used by Rudoy and Tserkovnikov (1975). For the three level system ($S = 1$), Loktev and Ostrovskii (1978), and Onufrieva (1981) have used SU(3) unitary transformations. In the case of arbitrary spin value SA magnets or in the general case of the multilevel Hamiltonian H_f, the problem of the exact diagonalization of H_f becomes nontrivial. A new approach (Val'kov 1988) to this problem is given below. There are two new moments in our approach in comparison with SU($2S + 1$) transformations mentioned above. Firstly, we write down the unitary transformation operators in atomic representation. Secondly, the generators of unitary rotations in the N-dimensional Hilbert space of the cell are given by the anti-Hermitian combinations of X-operators. It allows one to obtain simple and explicit form for the operators $U(\alpha)$ of the unitary transformation for an arbitrary dimension of the Hilbert space.

4.1 Unitary Operators and Transformation Laws in Atomic Representation

In a general case a single-ion Hamiltonian for the N-level system may be written in the following way

$$H = \sum_{p=1}^{N} \varepsilon_p X^{pp} + \sum_{\substack{p,q=1 \\ (p \neq q)}} V_{pq} X^{pq}, \quad V_{pq} = \overset{*}{V}_{qp}. \tag{4.1}$$

For spin Hamiltonians the states $|p\rangle$ used for the definition of X-operators in Eq. (4.1) may be taken as eigenstates of the S^z operator

$$S^z|p\rangle = (S - p + 1)|p\rangle, \quad p = 1, 2, \ldots, 2S + 1 = N. \tag{4.2}$$

The matrix elements V_{pq} determine the mixture of different states $|p\rangle$ in the eigenstates $|\Psi_p\rangle$ of the Hamiltonian (4.1):

$$H|\Psi_p\rangle = E|\Psi_p\rangle, \quad p = 1, 2, \ldots, N. \tag{4.3}$$

The eigenstate problem (4.3) may be reformulated as the problem of the construction of such unitary operator $\hat{O}$ that

$$|\Psi_p\rangle = \hat{O}^+|p\rangle, \quad p = 1, 2, \ldots, N.$$

The transformed Hamiltonian

$$\tilde{H} = \hat{O} H \hat{O}^+$$

becomes diagonal in the basis of the states $|p\rangle$. The unitary operator $\hat{O}$ may be written as

$$\hat{O} = \exp(\hat{B}),$$

with the anti-Hermitian operator $\hat{B}$. For a small dimension of the Hilbert space some special methods to construct the operator $\hat{O}$, for example using SU(3) group, are known but in the general case of arbitrary dimension this problem has not been solved. Here we show that it may be solved with the help of the X-operators.

We define the set of unitary operators

$$U_{nm}(\alpha) = \exp\left\{\alpha X^{nm} - \overset{*}{\alpha} X^{mn}\right\}, \quad m > n = 1, 2, \ldots, N. \tag{4.4}$$

This set contains $N(N-1)/2$ operators for an arbitrary value of the complex parameter α that may be specific for each X-operator (4.4), in other words $\alpha = \alpha_{mn}$. The unitary transformation operator $\hat{O}$ will be constructed using

$U_{nm}(\alpha)$ operators. The X-operator multiplication rule

$$X^{nm} X^{pq} = \delta_{mp} X^{nq} \qquad (4.5)$$

allows one to find explicitly the operator $U_{nm}(\alpha)$. We take the $U_{nm}(\alpha)$ operator in the form

$$U_{nm}(\alpha) = \sum_{p=0}^{\infty} (\alpha X^{nm} - \overset{*}{\alpha} X^{mn})^p \frac{1}{(p!)} = 1 + A_1 + \sum_{p=2}^{\infty} \frac{A_p}{(p!)}, \qquad (4.6)$$

where

$$A_1 = \alpha X^{nm} - \overset{*}{\alpha} X^{mn}, \quad A_p = (A_1)^p.$$

Using Eq. (4.5), it is easy to get the following expressions

$$A_2 = -|\alpha|^2 B, \quad B = X^{nn} + X^{mm},$$
$$A_3 = -|\alpha|^2 A_1,$$
$$A_4 = -(-1)^2 |\alpha|^4 B,$$
$$A_5 = -(-1)^2 |\alpha|^4 A_1,$$
$$\vdots$$

all of them may be written as

$$A_{2n} = -(-1)^n |\alpha|^{2n} B,$$
$$A_{2n+1} = -(-1)^n |\alpha|^{2n} A_1.$$

With the help of these relations, we obtain

$$U_{nm}(\alpha) = 1 + A_1 + B \sum_{k=1}^{\infty} \frac{(-1)^k |\alpha|^{2k}}{(2k)!} + A_1 \sum_{k=1}^{\infty} \frac{(-1)^k |\alpha|^{2k}}{(2k+1)!},$$

and finally in the form

$$U_{nm}(\alpha) = 1 + (\cos|\alpha| - 1) \cdot (X^{nn} + X^{mm}) + \sin|\alpha| \cdot (e^{i\mu} X^{nm} - e^{-i\mu} X^{mn}), \qquad (4.7)$$

where $|\alpha|$ and μ are given by

$$\alpha = |\alpha| \exp(i\mu).$$

Below we show that the set of $N(N-1)/2$ operators $U_{nm}(\alpha)$ with the different values of α for each (m,n) pair may be used to get the diagonal form of the transformed Hamiltonian

$$H = \sum_{p=1}^{N} E_p X^{pp}. \tag{4.8}$$

The X-operators transforms under unitary rotations in N^2-dimensional Lie algebra $AU(N)$:

$$X^{pp} \to U_{nm}(\alpha) X^{pq} U^{+}_{nm}(\alpha) \equiv X^{pq}(\alpha). \tag{4.9}$$

For the diagonal operators this transformation is given by

$$X^{nn}(\alpha) = \cos^2 |\alpha| \cdot X^{nn} + \sin^2 |\alpha| X^{mm} - \frac{1}{2} \sin 2|\alpha| (e^{i\mu} X^{nm} + \text{h.c.}), \tag{4.10}$$

$$X^{mm}(\alpha) = \cos^2 |\alpha| \cdot X^{mm} + \sin^2 |\alpha| X^{nn} + \frac{1}{2} \sin 2|\alpha| (e^{i\mu} X^{nm} + \text{h.c.}), \tag{4.11}$$

$$X^{pp}(\alpha) = X^{pp}, \quad p \neq n, m. \tag{4.12}$$

The nondiagonal X-operator transforms in the following way

$$X^{nm}(\alpha) = \cos^2 |\alpha| \cdot X^{nm} - \sin^2 |\alpha| e^{-i2\mu} X^{mn}$$
$$+ \frac{1}{2} \sin 2|\alpha| e^{-i\mu} (X^{nn} - X^{mm}), \tag{4.13}$$

$$X^{mn}(\alpha) = \cos^2 |\alpha| \cdot X^{mn} - \sin^2 |\alpha| e^{i2\mu} X^{nm}$$
$$+ \frac{1}{2} \sin 2|\alpha| e^{i\mu} (X^{nn} - X^{mm}), \tag{4.14}$$

$$\left. \begin{array}{l} X^{np}(\alpha) = \cos |\alpha| X^{np} - \sin |\alpha| e^{-i\mu} X^{mp} \\ X^{pn}(\alpha) = \cos |\alpha| X^{pn} - \sin |\alpha| e^{i\mu} X^{pm} \end{array} \right\} p \neq n, m, \tag{4.15}$$

$$\left. \begin{array}{l} X^{pm}(\alpha) = \cos |\alpha| X^{pm} + \sin |\alpha| e^{-i\mu} X^{pn} \\ X^{mp}(\alpha) = \cos |\alpha| X^{mp} + \sin |\alpha| e^{i\mu} X^{np} \end{array} \right\} p \neq n, m, \tag{4.16}$$

$$X^{pq}(\alpha) = X^{pq}, \quad p, q \neq n, m. \tag{4.17}$$

It is worthwhile mentioning two moments concerning transformation laws (4.10–4.17): they are written for arbitrary dimension of the Hilbert space N and they have selective character; only the part of operators with the given (m, n) are transforming leaving all the other operators X^{pq} with $p, q \neq n, m$ invariant. It simplifies greatly the diagonalization procedure especially if the Hamiltonian matrix has blocks corresponding to quasi-two level or quasi-three level subsystems.

Parameters of the Hamiltonian (4.1) before and after the unitary rotation in N^2-dimensional AU(N) algebra are connected by the following recurrent relations

$$H \to H' = U_{nm}(\alpha) H U_{nm}^+(\alpha) = \sum_{p=1}^{N} \varepsilon'_p X^{pp} + \sum_{\substack{p,q=1 \\ (p \neq q)}}^{N} V'_{pq} X^{pq}, \qquad (4.18)$$

where the transformed Hamiltonian parameters are given by

$$\varepsilon'_n = \frac{1}{2}[\varepsilon_n + \varepsilon_m + (\varepsilon_n - \varepsilon_m)\cos 2|\alpha| + \sin 2|\alpha| \cdot (V_{nm} e^{-i\mu} + \text{h.c.})],$$

$$\varepsilon'_m = \frac{1}{2}[\varepsilon_n + \varepsilon_m - (\varepsilon_n - \varepsilon_m)\cos 2|\alpha| - \sin 2|\alpha|(V_{nm} e^{-i\mu} + \text{h.c.})],$$

$$V'_{nm} e^{i\mu} = \frac{1}{2}\left[\frac{\varepsilon_m - \varepsilon_n}{2}\sin 2|\alpha| + V_{nm} e^{-i\mu}\cos^2|\alpha| - V_{nm}^* e^{i\mu}\sin 2|\alpha|\right],$$

$$V'_{np} = V_{np}\cos|\alpha| + V_{mp} e^{i\mu}\sin|\alpha|,$$

$$V'_{mp} = V_{pm}\cos|\alpha| - V_{pn} e^{i\mu}\sin|\alpha|,$$

$$\varepsilon'_p = \varepsilon_p, \quad V'_{pq} = V_{pq}, \qquad p, q \neq n, m. \qquad (4.19)$$

These recurrent relations allow one to diagonalize the cell Hamiltonian explicitly.

4.2 Diagonalization of Two-Level and Quasi-Two-Level Forms

For the two level system the diagonalization is obtained by one transformation $U_{12}(\alpha)$. Taking the nondiagonal part of the transformed Hamiltonian to be zero we find the equation for the complex parameter α:

$$\frac{1}{2}(\varepsilon_2 - \varepsilon_1)\sin 2|\alpha| + V_{12} e^{-i\mu}\cos^2|\alpha| - \overset{*}{V}_{12} e^{i\mu}\sin^2|\alpha| = 0. \qquad (4.20)$$

For the mixed matrix element V_{12} in Eq. (4.1) given by

$$V_{12} = |V_{12}|\exp(i\varphi_{12}), \qquad (4.21)$$

we get the following equations for $|\alpha|$ and μ.

$$\frac{1}{2}(\varepsilon_2 - \varepsilon_1)\sin 2|\alpha| + |V_{12}|\cos 2|\alpha|\cos(\varphi_{12} - \mu) = 0,$$
$$\sin(\varphi_{12} - \mu) = 0. \tag{4.22}$$

The solutions of Eq. (4.22) may be written as

$$\mu = \varphi_{12} + \kappa\pi, \quad \kappa = 0, 1 \tag{4.23}$$

where $|\alpha|$ is determined by the equation

$$\cos 2|\alpha| = \frac{(\varepsilon_2 - \varepsilon_1)}{\{(\varepsilon_2 - \varepsilon_1)^2 + 4|V_{12}|^2\}^{1/2}},$$
$$\sin 2|\alpha| = \frac{(-1)^{\kappa+1}2|V_{12}|}{\{(\varepsilon_2 - \varepsilon_1)^2 + 4|V_{12}|^2\}^{1/2}}. \tag{4.24}$$

The eigenvalues of the Hamiltonian (4.1) for the two level system have the standard form

$$E_{1,2} = (\varepsilon_2 + \varepsilon_1)/2 \mp \sqrt{(\varepsilon_2 - \varepsilon_1)^2/4 + |V_{12}|^2}, \tag{4.25}$$

while the eigenstates are given by

$$|\Psi_1\rangle = U_{12}^+(\alpha)|1\rangle = \cos|\alpha|\cdot|1\rangle + \sin|\alpha|e^{-i\mu}|2\rangle,$$
$$|\Psi_2\rangle = U_{12}^+(\alpha)|2\rangle = \cos|\alpha|\cdot|2\rangle - \sin|\alpha|e^{-i\mu}|1\rangle. \tag{4.26}$$

In the same manner we can diagonalize the quasi-two level form

$$\varepsilon_n X^{nn} + \varepsilon_m X^{mm} + V_{nm} X^{nm} + \overset{*}{V}_{nm} X^{mn} \tag{4.27}$$

of the N-level Hamiltonian where all matrix elements V_{np}, V_{pm} with $p \neq n, m$ are equal to zero. Due to the selective property of the transformation $U_{nm}(\alpha)$, it changes only the block (4.27) leaving all the other states invariant. It means that Eqs. (4.21)–(4.26) can be used to diagonalize the form (4.27) after the index substitution $1 \to n, 2 \to m$.

4.3 The Diagonalization of Three-Level Forms

In general, the three level block in the Hamiltonian (4.1) may be written as follows

$$H_{(3)} = \varepsilon_p X^{pp} + \varepsilon_q X^{qq} + \varepsilon_r X^{rr} + \{V_{pq} X^{pq} + V_{pr} X^{pr} + V_{qr} X^{qr} + \text{h.c.}\}. \tag{4.28}$$

The diagonalization of this Hamiltonian requires two steps. First, we transform it into the sum of quasi-two level and one level subsystems by combining the two unitary transformations $U_{pr}(\alpha)$ and $U_{pq}(\beta)$:

$$\begin{aligned} H_{(3)} &\to H'_{(3)} = U_{pr}(\alpha) H U^+_{pr}(\alpha), \\ H'_{(3)} &\to H''_{(3)} = U_{pq}(\beta) H' U^+_{pq}(\beta), \quad \beta \equiv |\beta| e^{i\nu}. \end{aligned} \tag{4.29}$$

The Hamiltonian $H''_{(3)}$ parameters are related to the $H'_{(3)}$ Hamiltonian parameters by the recurrent relations similar to Eq. (4.19) with the substitution in the notations

$$\begin{aligned} n &\to p, \quad m \to q, \quad p \to r, \quad |\alpha| \to |\beta|, \\ \mu &\to \nu, \quad \varepsilon'_n \to \varepsilon''_p, \quad \varepsilon'_m \to \varepsilon''_q. \end{aligned} \tag{4.30}$$

For the state $|p\rangle$ to be the eigenstate of $H''_{(3)}$, we have the system of two equations

$$V''_{pr}(\alpha, \beta) = 0, \quad V''_{pq}(\alpha, \beta) = 0, \tag{4.31}$$

that can be written explicitly in the form

$$\begin{aligned} V'_{pr} \cos|\beta| + V'_{qr} e^{i\nu} \sin|\beta| &= 0, \\ \left(\frac{\varepsilon'_q - \varepsilon'_p}{2}\right) \sin 2|\beta| + V'_{pq} e^{-i\nu} \cos^2|\beta| - \overset{*}{V'}_{pq} e^{i\nu} \sin^2|\beta| &= 0, \end{aligned} \tag{4.32}$$

where

$$\begin{aligned} V'_{pr} &= e^{i\mu} \left[\frac{\varepsilon_r - \varepsilon_p}{2} \sin 2|\alpha| + V_{pr} e^{-i\mu} \cos^2|\alpha| - \overset{*}{V}_{pr} e^{i\mu} \sin^2|\alpha| \right], \\ V'_{qr} &= V_{qr} \cos|\alpha| - V_{qp} e^{i\mu} \sin|\alpha|, \\ V'_{pq} &= V_{pq} \cos|\alpha| + V_{rq} e^{i\mu} \sin|\alpha|, \\ \varepsilon'_p &= \frac{1}{2} \left[\varepsilon_p + \varepsilon_r + (\varepsilon_p - \varepsilon_r) \cos 2|\alpha| + (V_{pr} e^{-i\mu} + \text{h.c.}) \sin 2|\alpha| \right], \\ \varepsilon'_r &= \frac{1}{2} \left[\varepsilon_p + \varepsilon_r - (\varepsilon_p - \varepsilon_r) \cos 2|\alpha| - (V_{pr} e^{-i\mu} + \text{h.c.}) \sin 2|\alpha| \right], \\ \varepsilon'_q &= \varepsilon_q. \end{aligned} \tag{4.33}$$

Therefore the Hamiltonian $H''_{(3)}$ has the form

$$H''_{(3)} = E_p X^{pp} + \varepsilon''_q X^{qq} + \varepsilon''_r X^{rr} + (V''_{qr} X^{qr} + \text{h.c.}), \tag{4.34}$$

with the eigenvalue E_p given by

$$E_p = \frac{1}{2}(\varepsilon'_p + \varepsilon'_q) - \frac{1}{2}\{(\varepsilon'_p + \varepsilon'_q)^2 + 4|V'_{pq}|^2\}^{1/2}. \tag{4.35}$$

The parameters of the quasi-two level form in the Hamiltonian (4.34) are equal to

$$\begin{aligned}\varepsilon''_q &= \frac{1}{2}(\varepsilon'_p + \varepsilon'_q) + \frac{1}{2}\{(\varepsilon'_p - \varepsilon'_q)^2 + 4|V'_{pq}|^2\}^{1/2}, \\ \varepsilon''_r &= \varepsilon'_r, \quad V''_{qr} = V'_{qr} \cos|\beta| - V'_{pr} e^{-i\nu} \sin|\beta|.\end{aligned} \tag{4.36}$$

Choosing the argument of the parameter β in the form

$$\nu = \varphi'_{pq} + \pi,$$

where φ'_{pq} is an argument of the complex value

$$V'_{pq} = |V'_{pq}| \exp(i\varphi'_{pq}),$$

we get the modulus of β:

$$\begin{aligned}\cos 2|\beta| &= \frac{(\varepsilon'_q - \varepsilon'_p)}{\{(\varepsilon'_q - \varepsilon'_p)^2 + 4|V'_{pq}|^2\}^{1/2}}, \\ tg2|\beta| &= \frac{2|V'_{pq}|}{(\varepsilon'_q - \varepsilon'_p)}.\end{aligned} \tag{4.37}$$

In the second step, the quasi-two level form in the Hamiltonian (4.34) is diagonalized by the unitary transformation $U_{qr}(\gamma)$, where $\gamma = |\gamma| \exp(i\delta)$, using the method described in the previous section. Thus results in the diagonal form

$$H'''_{(3)} = U_{qr}(\gamma) H''_{(3)} U^+_{qr}(\gamma) = E_p X^{pp} + E_q X^{qq} + E_r X^{rr}, \tag{4.38}$$

where the eigenenergies are given by

$$\begin{aligned}E_q &= \frac{\varepsilon''_r + \varepsilon''_q}{2} - \left\{\left(\frac{\varepsilon''_r - \varepsilon''_q}{2}\right)^2 + |V''_{qr}|^2\right\}^{1/2}, \\ E_r &= \frac{\varepsilon''_r + \varepsilon''_q}{2} + \left\{\left(\frac{\varepsilon''_r - \varepsilon''_q}{2}\right)^2 + |V''_{qr}|^2\right\}^{1/2}.\end{aligned} \tag{4.39}$$

The corresponding eigenvectors are equal to

$$|\Psi_P\rangle = U_{pr}^{-1}(\alpha)U_{pq}^{-1}(\beta)|p\rangle$$
$$= \cos|\beta|\cos|\alpha| \cdot |p\rangle + \sin|\beta|e^{-i\nu}|q\rangle + \sin|\alpha|\cos|\beta|e^{-i\mu}|r\rangle, \quad (4.40)$$
$$|\Psi_q\rangle = U_{pr}^{-1}(\alpha)U_{pq}^{-1}(\beta)U_{qr}^{-1}(\gamma)|q\rangle$$
$$= \cos|\gamma|\cos|\beta| \cdot |q\rangle$$
$$+ \left\{\cos|\alpha|\sin|\gamma|e^{-i\delta} - \sin|\alpha| \cdot \sin|\beta|\cos|\gamma|e^{i(\nu-\mu)}\right\}|r\rangle$$
$$- \left\{\sin|\alpha|\sin|\gamma|e^{i(\mu-\delta)} + \cos|\alpha| \cdot \sin|\beta| \cdot \cos|\gamma|e^{i\nu}\right\}|p\rangle, \quad (4.41)$$
$$|\Psi_r\rangle = U_{pr}^{-1}(\alpha)U_{pq}^{-1}(\beta)U_{qr}^{-1}(\gamma)|r\rangle$$
$$= -\sin|\gamma|\cos|\beta|e^{i\delta}|q\rangle$$
$$+ \left\{\cos|\alpha| \cdot \cos|\gamma| + \sin|\alpha|\sin|\beta| \cdot \sin|\gamma|e^{i(\nu+\delta-\mu)}\right\}|r\rangle$$
$$+ \left\{\cos|\alpha| \cdot \sin|\beta| \cdot \sin|\gamma\rangle e^{i(\nu+\delta)} - \sin|\alpha|\cos|\gamma|e^{i\mu}\right\}|p\rangle. \quad (4.42)$$

The explicit form of the eigenvectors (4.40)–(4.42) is especially convenient when some self-consistent fields are included in the cell Hamiltonian (for example, the magnetization or the mean values of the quadrupolar moments). To illustrate this statement we consider the model of an easy plane anisotropic Heisenberg ferromagnet in the external magnetic field oriented in the easy magnetization plane. The Hamiltonian is

$$H = -\frac{1}{2}\sum_{fm} I_{fm}(\boldsymbol{S}_f \boldsymbol{S}_m) + \sum_f 2D(S_f^x)^2 - H\sum_f S_f^z. \quad (4.43)$$

Here the interatomic exchange interaction I_{fm} is responsible for long-range magnetic order, the second term being the SA of the easy plane type with $D > 0$ and the $0x$ axis oriented along the anisotropy direction. The external magnetic field in the last Zeeman term is given in energy units, $g\mu_B H \to H$. The $0z$ axis lies in the easy magnetization plane.

Intra-atomic energy levels of the Hamiltonian (4.43) in the approximation of the anisotropic molecular field are given by the operator

$$H_0(f) = \frac{1}{2}I_0\sigma^2 + 2D(S_f^x)^2 - \bar{H}S_f^z, \quad (4.44)$$

that incorporates all intra-atomic correlations (the anisotropy in this case) exactly. The self-consistent parameter $\sigma = \langle S^z \rangle$ determines an effective field

$$\bar{H} = H + I_0\sigma, \quad (4.45)$$

where I_0 is the $q = 0$ component of an exchange parameter's Fourier transform.

We denote the eigenvectors of the operator S_j^z by

$$S^z|n\rangle = (S - n + 1)|n\rangle, \quad n = 1, 2, \ldots, 2S + 1, \qquad (4.46)$$

and construct the X-operators on this basis

$$X^{pq} \equiv |p\rangle\langle q|. \qquad (4.47)$$

The action of the spin operator results in the following expressions

$$S^+|n\rangle = \sqrt{(n-1)(2S+2-n)}|n-1\rangle,$$
$$S^-|n\rangle = \sqrt{n(2S-n+1)}|n+1\rangle, \qquad (4.48)$$

that are useful for the X-operator representation of the Hamiltonian (4.44). Here we present it for the spin values $S = 1, 3/2, 2$:

(i) For $S = 1$, the Hamiltonian (4.44) is given by

$$H_0 = \frac{1}{2}I_0\sigma^2 + D + DX^{22} - \bar{H}(X^{11} - X^{33}) + D(X^{13} + X^{31}). \qquad (4.49)$$

(ii) For $S = 3/2$, the Hamiltonian H_0 can be written as

$$H_0 = \frac{1}{2}I_0\sigma^2 + \frac{3}{2}D - \frac{3\bar{H}}{2}X^{11} + \left(2D + \frac{\bar{H}}{2}\right)X^{33} + \sqrt{3}D(X^{13} + X^{31})$$
$$+ \left(2D - \frac{\bar{H}}{2}\right)X^{22} + \frac{3\bar{H}}{2}X^{44} + \sqrt{3}D(X^{24} + X^{42}). \qquad (4.50)$$

(iii) For $S = 2$, the structure of the Hamiltonian H_0 is

$$H_0 = \frac{1}{2}I_0\sigma^2 + 2D - (\bar{H} - 3D)X^{22} + (\bar{H} + 3D)X^{44} + 3D(X^{24} + X^{42})$$
$$+ 2\bar{H}(X^{11} - X^{55}) + 4DX^{33} + \sqrt{6}D(X^{13} + X^{31} + X^{35} + X^{53}). \qquad (4.51)$$

In all examples the Hamiltonian H_0 is the sum of quasi-two and quasi-three level subsystems. For $S = 1$ the state $|2\rangle$ is eigenstate and the task of diagonalization of the Hamiltonian (4.49) is reduced to the diagonalization of the quasi-two level subsystem given by

$$H_{(2)} = -\bar{H}(X^{11} - X^{33}) + D(X^{13} + X^{31}). \qquad (4.52)$$

In this case, vector $|2\rangle$ is an eigenvector for the transformed Hamiltonian $H_0' = U_{13} H U_{13}^+$. In the ferromagnetic phase the ground state at $T = 0$ is described by the eigenstate $|\Psi_1\rangle$ which is equal to

$$|\Psi_1\rangle = U_{13}^+(\alpha)|1\rangle = \cos\alpha|1\rangle + \sin\alpha|-1\rangle. \qquad (4.53)$$

Due to real values of the parameters in Eq. (4.52), the parameter α is also real. The magnetization σ is given by

$$\sigma = \cos 2\alpha \qquad (4.54)$$

and Eq. (4.20) has the form

$$(H + I_0 \cos 2\alpha)\sin 2\alpha + D\cos 2\alpha = 0. \qquad (4.55)$$

Thus, the self-consistent field has been expressed via the unitary rotation parameter α by Eq. (4.54). Equation (4.55) is in a close form and its solution gives the value of the parameter α.

In the absence of an external field, $H = 0$, these are the following possible phases:

(1) Quadrupolar phase with $\sigma = 0$, $\cos 2\alpha = 0$.
(2) Magnetic phase at $D < I_0$ with $\sigma \neq 0$, $\sin 2\alpha = -D/I_0$.

At $D > I_0$, the energy E_F in the magnetic phase is less than the quadrupolar phase energy E_Q. In the magnetic phase

$$\sigma = \sqrt{1 - (D/I_0)^2}, \quad E_F = -\frac{1}{2}D - \frac{D^2}{2I_0} < 0.$$

For $D > I_0$, only one solution for the quadrupolar phase is possible. The point $D = I_0$ is the transition point from ferromagnetic to quadrupolar phase.

For the spin $S = 3/2$, the single ion Hamiltonian is the sum of two quasi-two level forms and is diagonalized by two unitary rotations $U(\alpha, \beta) = U_{13}(\alpha)U_{24}(\beta)$.

Of greater interest is the case of $S = 2$, where the ground state and the magnetic characteristics are determined by the three level form

$$H_{(3)} = \frac{1}{2}I_0\sigma^2 + \varepsilon_5(X^{55} - X^{11}) + \varepsilon_3 X^{33} + V(X^{13} + X^{35} + \text{h.c.}), \qquad (4.56)$$

with

$$\varepsilon_5 = 2\bar{H}, \quad \varepsilon_3 = 4D, \quad V = \sqrt{6D}.$$

The diagonalization of the three-level form has been described above. At first we do two unitary rotations $U_{15}(\alpha)$ and $U_{13}(\beta)$ with the real parameters α and β because the matrix elements in Eq. (4.56) are real. The system of equations for α and β is written in the form

$$2H \sin 2\alpha \cos \beta + \sqrt{6}D(\cos\alpha - \sin\alpha)\sin\beta = 0,$$
$$(2D + \bar{H}\cos 2\alpha)\sin 2\beta + \sqrt{6}D(\cos\alpha + \sin\alpha)\cos 2\beta = 0. \quad (4.57)$$

The ground state function is given by

$$|\Psi_1\rangle = U_{15}^+(\alpha)U_{13}^+(\beta)|1\rangle = \cos\alpha\cos\beta|1\rangle + \sin\beta|3\rangle + \sin\alpha\cos\beta|5\rangle. \quad (4.58)$$

At $T = 0$, the magnetization is equal to

$$\sigma = \langle\Psi_1|S^z|\Psi_1\rangle = 2\cos^2\beta\cos 2\alpha. \quad (4.59)$$

Taking into account the relation of ε_5 to σ and to the α, β parameters

$$\varepsilon_5 = 2\bar{H} = 2H + 4I_0\cos^2\beta\cos 2\alpha,$$

we have the closed system of Eqs. (4.57) for α and β. The dependence of the solutions of this equations on the anisotropy value at $H = 0$ is shown in Fig. 4.1. At $D = 3I_0$, when $\alpha = \pi/4$ we obtain $\sigma = 0$. This is the point of the ferromagnetic–quadrupolar transition. Increasing D results in increasing values of α and β and decreasing magnetization.

In the external magnetic field, the magnetization σ is non-zero for all values of the parameter D/I_0 (Fig. 4.2). The energy levels obtained by the

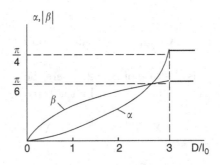

Fig. 4.1 The anisotropy dependence of the unitary transformation parameters α and β at $H = 0$ for the $S = 2$ SA magnet.

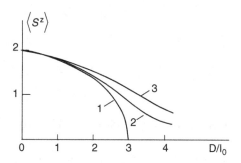

Fig. 4.2 The dependence of the magnetization on the anisotropy value D/I_0 for the easy plane ferromagnet at $T = 0$ and 3 different values of the magnetic field $H = 0(1)$, $H/I_0 = 0.1(2)$, $H/I_0 = 0.3(3)$.

exact diagonalization of the Hamiltonian (4.51) are given by

$$E_1 = 2D - \bar{H}\cos 2\alpha - \sqrt{(2D + \bar{H}\cos 2\alpha)^2 + 6D^2(1 + \sin 2\alpha)},$$
$$E_2 = 5D - \sqrt{\bar{H}^2 + 9D^2}, \quad E_4 = 5D + \sqrt{\bar{H}^2 + 9D^2},$$
$$E_3 = 2D + \frac{\varepsilon_3'' + \varepsilon_5'}{2} - \sqrt{\left(\frac{\varepsilon_5' - \varepsilon_3''}{2}\right) + (V_{35}'')^2}, \qquad (4.60)$$
$$E_5 = 2D + \frac{\varepsilon_3'' + \varepsilon_5'}{2} + \sqrt{\left(\frac{\varepsilon_5' - \varepsilon_3''}{2}\right) + (V_{35}'')^2},$$

where

$$\varepsilon_5' = 2\bar{H}\cos 2\alpha, \quad V_{35}'' = \sqrt{6}D(\cos\alpha - \sin\alpha)\cos\beta - 2\bar{H}\sin 2\alpha \sin\beta,$$
$$\varepsilon_5'' = 2D - \bar{H}\cos 2\alpha + \sqrt{(2D + \bar{H}\cos 2\alpha)^2 + 6D^2(1 + \sin 2\alpha)}.$$

Besides having the ground state eigenvector (4.58), excited states eigenvectors are equal to

$$\begin{aligned}|\Psi_2\rangle &= \cos\Omega\,|2\rangle - \sin\Omega\,|4\rangle,\\ |\Psi_3\rangle &= \cos\gamma\cos\beta\,|3\rangle + (\cos\alpha\sin\gamma - \sin\alpha\sin\beta\cos\gamma)\,|5\rangle\\ &\quad - (\sin\alpha\sin\gamma + \cos\alpha\sin\beta\cos\gamma)\,|1\rangle,\\ |\Psi_4\rangle &= \sin\Omega\,|2\rangle + \cos\Omega\,|4\rangle,\\ |\Psi_5\rangle &= (\cos\alpha\sin\beta\sin\gamma - \sin\alpha\cos\gamma)\,|1\rangle\\ &\quad + (\cos\alpha\cos\gamma + \sin\alpha\sin\beta\sin\gamma)\,|5\rangle - \sin\gamma\cos\beta\,|3\rangle,\end{aligned} \qquad (4.61)$$

where the parameters Ω and γ are determined as follows

$$\cos 2\Omega = \bar{H}/\sqrt{\bar{H}^2 + 9D^2}, \qquad \sin 2\Omega = 3D/\sqrt{\bar{H}^2 + 9D^2},$$

$$\cos 2\gamma = \frac{\varepsilon_5' - \varepsilon_3''}{\sqrt{(\varepsilon_5' - \varepsilon_3'')^2 + 4(V_{35}'')^2}}, \qquad \sin 2\gamma = \frac{2V_{35}''}{\sqrt{(\varepsilon_5' - \varepsilon_3'')^2 + 4(V_{35}'')^2}}.$$

4.4 The Diagonalization of the N Level Hamiltonian

The method of unitary transformations described above can be easily extended to the N-level systems. By $(N-1)$ subsequent unitary transformations (4.18) given by

$$\begin{aligned} H \to H' &= U_{1N}(\alpha) H U_{1N}^+(\alpha), \\ H' \to H'' &= U_{1,N-1}(\beta) H' U_{1,N-1}^+(\beta), \\ &\vdots \\ H^{\overbrace{''\cdots'}^{N-2}} \to H^{\overbrace{''\cdots'}^{N-1}} &= U_{1,2}(\xi) H^{\overbrace{''\cdots'}^{N-2}} U_{12}^+(\xi), \end{aligned} \qquad (4.62)$$

we can write the Hamiltonian $H^{\overbrace{''\cdots'}^{N-1}}$ in the form

$$H^{\overbrace{''\cdots'}^{N-1}} = E_1 X^{11} + \sum_{p=2}^{N} \varepsilon_p^{\overbrace{''\cdots'}^{N-1}} X^{pp} + \sum_{\substack{p,q=2 \\ (p \neq q)}}^{N} V_{pq}^{\overbrace{''\cdots'}^{N-1}} \cdot X^{pq} \qquad (4.63)$$

which is the direct sum of the diagonal operator $E_1 X^{11}$ and the orthogonal $(N-1)$-level operator if the $(N-1)$ parameters $\alpha, \beta, \ldots, \xi$ satisfy the equations

$$V_{1N}^{\overbrace{''\cdots'}^{N-1}}(\alpha, \beta, \ldots, \xi) = 0, \ldots, \qquad V_{12}^{\overbrace{''\cdots'}^{N-1}}(\alpha, \beta, \ldots, \xi) = 0. \qquad (4.64)$$

The energy E_1 is given by Eq. (4.35) with the indexes substitution $p \to 1, q \to 2$, and with all parameters in the right part of Eq. (4.35) having $(N-2)$ primes. Using procedure applied to the $(N-1)$ level form, we obtain the eigenstate $|\Psi_2\rangle$ with energy E_2 and the remaining $(N-2)$-level form, and etc., up to the final diagonalization of the single ion Hamiltonian.

Chapter 5

Diagram Technique in the X-Operators Representation

This technique is not so simple unlike the Fermi and Bose operators, due to the more complicate commutation rules (1.42). Indeed the commutator (anticommutator) of the Bose (Fermi) operators is a c-number so the pairing of two operators under the symbol of averaging (contraction) in the perturbation expansion decreases the number of operators by two, resulting in the Wick's theorem. For the X-operators the commutator (anticommutator) is a new X-operator, so the pairing of two operators decreases the number of operators by one. Using Gaudin (1960) ideas on recurrent relations, Westwanski and Pavlikovski (1973) had proved the Wick's theorem for X-operators. This theorem is similar to the Wick's theorem for spin operators (Vaks et al 1967a, 1967b; Izyumov and Kassan-Ogly 1970a, 1970b; Izyumov et al 1974). Several groups have proposed different variants of the diagram technique for X-operators (Slolodyan and Stasyuk 1974; Barabanov et al 1974, 1975) but all of them was rather cumbersome.

The decisive step has been done by Zaitsev (1975, 1976) who has introduced a notion of root vectors $\alpha(p,q)$ that simplifies greatly the diagrams. The construction of the Cartan–Weyl local basis with spurless diagonal operators used by Zaitsev can be easily done for systems with a small number of local states. In the Hubbard model this number is equal to 4. In the easy-axis Heisenberg ferromagnet with $S = 1$ and in the t–J model there are 3 local states. When the number of local states is large Zaitsev's method becomes inconvenient. Moreover if this number is arbitrary and is not fixed Zaitsev's method cannot be realized in principle. Sandalov and Podmarkov (1982) have shown that the advantage of introducing root vectors is conserved if diagonal operators of Weyl basis are used. A step forward in this technique was made by Val'kov and Ovchinnikov (1983) who represented components of root vectors $\alpha(p,q)$ as follows

$$\alpha_i(p,q) = \delta_{ip} - \delta_{iq}.$$

It is essentially that representation is valid for arbitrary number of local states.

One very serious drawback of X-operators diagram technique was found by Vedyaev and Nikolaev (1984). They have shown that one and the same average of several X-operators can be expressed by different set of diagrams depending on the way of pairing in the Wick's theorem. To avoid this ambiguity they have proposed the hierarchy principle for the nondiagonal operators. With this principle the diagram technique for X-operators in the Hubbard model and the t–J model is described in detail in several books (Izyumov and Skryabin 1987; Izyumov et al 1994). Again when the number of local states is large or arbitrary the hierarchy principle becomes very inconvenient. That is why we give here a new formulation of diagram technique for X-operators that is valid for arbitrary number of local states and can be applied to the models described in the Chapter 2. To avoid the ambiguity problem we introduce the notion of a generating operator, a principle of topological continuity and a hierarchy of the interaction lines (see below).

5.1 Green's Function in the X-Operators Representation

We consider a system of particles with the two-particle interactions given by the Hamiltonian

$$H = \sum_f \hat{H}(f) + \frac{1}{2}\sum_{fg}\hat{H}(f,g). \tag{5.1}$$

Here the first term describes a set of the non-interacting unit cells (while the intracell strong correlations are included in $H(f)$). The solution of the eigenstate problem

$$\hat{H}(f)|\Psi_n(f)\rangle = E_n|\Psi_n(f)\rangle,$$

results in the Hubbard X-operators

$$X_f^{nm} = |\Psi_n(f)\rangle\langle\Psi_m(f)|.$$

In the X-representation the Hamiltonian (5.1) has the form

$$H = H_0 + H_{\text{int}} \qquad (5.2)$$

with the diagonal H_0 term:

$$H_0 = \sum_f \sum_n E_n X_f^{nn} \equiv \sum_{fn} E_n h_{fn}, \quad h_{fn} \equiv X_f^{nn}. \qquad (5.3)$$

The intercell interaction consists of three terms

$$H_{\text{int}} = \frac{1}{2} \sum_{fg} \sum_{\alpha\beta} V_{fg}^{\alpha\beta} X_f^\alpha X_g^\beta + \sum_{fg} \sum_{\alpha n} V_{fg}^{\alpha n} X_f^\alpha X_g^{nn}$$

$$+ \frac{1}{2} \sum_{fg} \sum_{nm} V_{fg}^{nm} X_f^{nn} X_g^{mm}, \qquad (5.4)$$

that we call transverse, transverse–longitudinal and longitudinal interactions, correspondingly. For example in the Hubbard model where the perturbation is induced by hopping there is only the transverse interaction. In the t–J model, both transverse and longitudinal interactions exist. In general the mixed transverse–longitudinal term is also possible.

To study a quasiparticle energy spectrum of a system and its thermodynamics we introduce as usually in the many-body physics finite temperature Green's functions in Matsubara representation (Abrikosov *et al* 1965):

$$D_{\alpha\beta}(f\tau; g\tau') = -\langle T_\tau \tilde{X}_f^\alpha(\tau) \tilde{X}_g^{-\beta}(\tau') \rangle,$$
$$D_{\alpha m}(f\tau; g\tau') = -\langle T_\tau \tilde{X}_f^\alpha(\tau) \tilde{X}_g^{mm}(\tau') \rangle, \qquad (5.5)$$
$$D_{nm}(f\tau; g\tau') = -\langle T_\tau \tilde{X}_f^{nn}(\tau) \tilde{X}_g^{mm}(\tau') \rangle.$$

In our case the Green's functions are determined via X-operators in the "Heisenberg representation" with Matsubara time τ:

$$\tilde{X}_f^\alpha(\tau) = e^{\tau H} X_f e^{-\tau H}, \quad \tilde{X}_f^{nn}(\tau) = e^{\tau H} X_f^{nn} e^{-\tau H}.$$

In Eq. (5.5), T_τ is the time ordering operator

$$T_\tau \tilde{A}(\tau) \tilde{B}(\tau') = \begin{cases} \tilde{A}(\tau)\tilde{B}(\tau') & \tau > \tau' \\ \pm \tilde{B}(\tau')\tilde{A}(\tau) & \tau < \tau' \end{cases}.$$

At $\tau < \tau'$ the lower sign corresponds to both Fermi type operators, in all other cases the upper sign is used. The averaging in Eq. (5.5) is done with a total density matrix

$$\langle T_\tau \tilde{A}(\tau)\tilde{B}(\tau')\rangle = \frac{Sp(T_\tau \tilde{A}(\tau)\tilde{B}(\tau')e^{-H/T})}{Sp(e^{-H/T})}, \quad (5.6)$$

where T is a temperature in the energy units ($k_B = 1$). The Matsubara time lies in a $(0, 1/T)$ interval. It is clear from Eq. (5.6) that the Green's functions depend only on the time difference $\tau - \tau'$. Indeed, for $\tau > \tau'$ we get

$$Sp(T_\tau \tilde{A}(\tau)\tilde{B}(\tau')e^{-H/T}) = Sp\{e^{\tau H} A e^{-(\tau-\tau')H} B e^{-\tau' H} e^{-H/T}\}$$
$$= Sp\{e^{(\tau-\tau')H} A e^{-(\tau-\tau')H} B e^{-H/T}\}$$
$$= Sp\{\tilde{A}(\tau - \tau')B e^{-H/T}\}.$$

Here we have used the cyclic permutation symmetry under the spur and the commutation of exponential operators $\exp(-\tau' H)$ and $\exp(-H/T)$. The same $\tau - \tau'$ dependence appears also for $\tau < \tau'$, meaning that the Green's functions depend on $\tau - \tau'$ in the interval $(-1/T, 1/T)$.

Thus, in Fourier transformation, a Green's function is periodically continuated from the interval $(-1/T, 1/T)$ to the infinite τ axis. That is why a frequency in the Fourier series is discrete, $\omega_n = \pi n T, n = 0, \pm 1, \pm \ldots$. For negative T: $-1/T < \tau < 0$,

$$-\langle T_\tau A(\tau)B\rangle = \mp \langle BA(\tau)\rangle = \mp \langle A(\tau + 1/T)B\rangle.$$

This formula gives the relation between Green's functions with positive and negative τ

$$D_{\lambda\lambda'}(\tau < 0) = \pm D_{\lambda\lambda'}(\tau + 1/T).$$

Here again the lower sign is used if and only if both operators in the Green's functions are of Fermi type. It is easy to prove that non-zero components in the Fourier transformation

$$D_{\lambda\lambda'}(f\tau, g\tau') = \frac{T}{N} \sum_{\mathbf{k},\omega_n} \exp\{i\mathbf{k} \cdot (\mathbf{R}_f - \mathbf{R}_g) - i\omega_n(\tau - \tau')\} D_{\lambda\lambda'}(\mathbf{k}, \omega_n)$$
(5.7)

have only odd frequencies for the Fermi type Green's function

$$\omega_n = (2n+1)\pi T, \quad n = 0, \pm 1, \pm 2, \ldots,$$

and only even frequencies for the Bose type Green's function

$$\omega_n = 2\pi n T, \quad n = 0, \pm 1, \pm 2, \ldots.$$

In the perturbation theory the time dependence of all operators is determined by H_0 in the interaction representation (Abrikosov et al 1965):

$$A(\tau) = e^{\tau H_0} A e^{-\tau H_0}.$$

The Heisenberg representation and the interaction one are connected by the $\hat{S}$-matrix

$$e^{-\tau H} = e^{-\tau H_0} \hat{S}(\tau),$$

that satisfies to the equation

$$\frac{d\hat{S}}{d\tau} = -H_{\text{int}}(\tau)\hat{S}(\tau), \tag{5.8}$$

where $H_{\text{int}}(\tau)$ is written in the interaction representation as

$$H_{\text{int}}(\tau) = e^{\tau H_0} H_{\text{int}} e^{-\tau H_0}.$$

A solution of Eq. (5.8) may be written in the form

$$\hat{S}(\tau) = T_\tau \exp\left\{-\int_0^\tau H_{\text{int}}(\tau_1) d\tau_1\right\}.$$

The matrix $\hat{S}(1/T)$ with $\tau = 1/T$ is called the finite temperature scattering matrix and is used to calculate the thermodynamic averages:

$$Sp(e^{-H/T}) = Sp\{\hat{S}(1/T)e^{-H_0/T}\} = \langle \hat{S}(1/T)\rangle_0 \cdot Sp\left(e^{-H_0/T}\right), \tag{5.9}$$

where an angular brackets with subscript "0" mean the averaging with H_0:

$$\langle \hat{A}\rangle_0 = Sp(\hat{A} e^{-H_0/T})/Sp(e^{-H_0/T}).$$

From the numerator of the Green's function at $\tau > \tau'$, there are several obvious identities

$$Sp\{T_\tau \tilde{X}_f^\lambda(\tau) \tilde{X}_g^{\lambda'}(\tau') e^{-H/T}\}$$
$$= Sp\{e^{\tau H} X_f^\lambda e^{-\tau H} e^{\tau' H} X_g^{\lambda'} e^{-\tau' H} e^{-H/T}\}$$
$$= Sp\{\hat{S}^{-1}(\tau) X_f^\lambda(\tau) \hat{S}(\tau) \hat{S}^{-1}(\tau') X_g^{\lambda'}(\tau') e^{-H_0/T} \hat{S}(1/T)\}$$
$$= Sp\{\hat{S}(1/T,\tau) X_f^\lambda(\tau) \hat{S}(\tau,\tau') X_g^{\lambda'}(\tau') \hat{S}(\tau',0) e^{-H_0/T}\}$$
$$= Sp\{[T_\tau X_f^\lambda(\tau) X_g^{\lambda'}(\tau') \hat{S}(1/T)] e^{-H_0/T}\}.$$

Here we introduce a matrix of evolution

$$\hat{S}(\tau_2,\tau_1) = T_\tau \exp\left\{-\int_{\tau_1}^{\tau_2} H_{\text{int}}(\tau)d\tau_1\right\}, \quad \tau_2 > \tau_1$$

with the properties

$$\hat{S}(\tau_3,\tau_2)\hat{S}(\tau_2,\tau_1) = \hat{S}(\tau_3,\tau_1), \quad \tau_3 > \tau_2 > \tau_1,$$

$$\hat{S}(\tau) = \hat{S}(\tau,0),$$

$$\hat{S}(\tau_2,\tau_1) = \hat{S}(\tau_2)S^{-1}(\tau_1).$$

At last the Green's function in the interaction representation is given by

$$D_{\lambda\lambda'}(f\tau;g\tau') = -\frac{\langle T_\tau X_f^\lambda(\tau) X_g^{\lambda'}(\tau)\hat{S}(1/T)\rangle_0}{\langle \hat{S}(1/T)\rangle_0}, \qquad (5.10)$$

and can be calculated by the perturbation expansion

$$D_{\lambda\lambda'}(f\tau;g\tau') = -\frac{1}{\langle \hat{S}(1/T)\rangle_0} \sum_{n=0}^{\infty} \frac{(-1)^n}{n!} \int_0^{1/T} d\tau_1 \int_0^{1/T} d\tau_2 \cdots \int_0^{1/T} dt_n$$

$$\times \langle T_\tau X_f^\lambda(\tau) X_g^{\lambda'}(\tau') H_{\text{int}}(\tau_1) H_{\text{int}}(\tau_2) \cdots H_{\text{int}}(\tau_n)\rangle_0. \qquad (5.11)$$

Since $H_{\text{int}}(\tau)$ is a bilinear form of X-operators, to obtain the n-th order contribution to Eq. (5.11) one have to calculate the average of the time ordered product of $2n+2$ X-operators with both diagonal and nondiagonal X-operators

$$\langle T_\tau X_f^\alpha(\tau) X_g^{-\beta}(\tau') X_e^{nn}(\tau_1) \cdots X_r^\gamma(\tau_n)\rangle_0. \qquad (5.12)$$

This can be done with the help of the Wick's theorem.

5.2 Wick's Theorem for the Hubbard Operators

We consider the term

$$A_n = -\langle T_\tau X_1^{\lambda_1}(\tau_1) X_2^{\lambda_2}(\tau_2) \cdots X_{k-1}^{\lambda_{k-1}}(\tau_{k-1}) X_f^\alpha(\tau) X_{k+1}^{\lambda_{k+1}}(\tau_{k+1}) \cdots$$

$$X_n^{\lambda_n}(\tau_n)\rangle_0, \qquad (5.13)$$

where $X_i^{\lambda_i}(\tau_i)$ is the diagonal operator $X_i^{nn}(\tau_i) = X_i^{nn}$ if λ_i is a diagonal index (n,n), otherwise $X_i^{pq}(\tau_i)$ is the nondiagonal operator $X_f^\alpha(\tau)$ when $\alpha_i = (p,q)$. Calculation starts with the nondiagonal operator $X_f^\alpha(\tau)$ that is in k-th position in Eq. (5.13), we mark it with a special notation similar to the Wick's theorem for the spin operators (Izyumov et al 1974). We call this operator the generating operator meaning that it starts to generate the pairing of two X-operators (see below).

Without the loss of generality we can assume that a set of X-operators in Eq. (5.13) is time ordered otherwise we should permute operators in the ordered position and to count the sign determining by the number of Fermi operator permutations. Then we can give up the T_τ operator in Eq. (5.13).

When the generating operator is of the Fermi type then we can write A_n in the form

$$A_n = -\sum_{i=k+1}^{n} (-1)^{p_i} \langle X_1^{\lambda_1}(\tau_1) \cdots X_{k-1}^{\lambda_{k-1}}(\tau_{k-1}) X_{k+1}^{\lambda_{k+1}}(\tau_{k+1}) \cdots$$
$$\times [X_f^\alpha(\tau) X_i^{\lambda_i}(\tau_i)]_\pm \cdots X_n^{\lambda_n}(\tau_n)\rangle_0$$
$$\pm (-1)^{p_n} \langle X_1^{\lambda_1}(\tau_1) \cdots X_{k-1}^{\lambda_{k-1}}(\tau_{k-1}) X_{k+1}^{\lambda_{k+1}}(\tau_{k+1}) \cdots X_n^{\lambda_n}(\tau_n) X_f^\alpha(\tau)\rangle_0,$$
(5.14)

where $[X_f^\alpha(\tau), X_i^{\lambda_i}(\tau_i)]_\pm$ is the anticommutator (commutator) if $X_i^{\lambda_i}$ is the Fermi (Bose) type operator, p_i is number of $X_f^\alpha(\tau)$ operator permutations with the Fermi type operators from the initial position in Eq. (5.13) to the position near and left to the operator $X_i^{\lambda_i}(\tau_i)$. A sign ($\pm$) before the last term in Eq. (5.14) is chosen $+(-)$ if the operator $X_n^{\lambda_n}(\tau_n)$ is of the Fermi (Bose) type. The last term in Eq. (5.14):

$$B_n \equiv \langle X_1^{\lambda_1}(\tau_1) \cdots X_{k-1}^{\lambda_{k-1}}(\tau_{k-1}) X_{k+1}^{\lambda_{k+1}}(\tau_{k+1}) \cdots X_n^{\lambda_n}(\tau_n) X_f^\alpha(\tau)\rangle_0,$$

can be evaluated in the following way ($Z_0 = Sp\{\exp(-H_0/T)\}$).

$$B_n = \frac{1}{Z_0} Sp\{X_1^{\lambda_1}(\tau_1) \cdots X_n^{\lambda_n}(\tau_n) X_f^\alpha(\tau) e^{-H_0/T}\}$$
$$= \frac{1}{Z_0} Sp\left\{X_f^\alpha(\tau) e^{-H_0/T} X_1^{\lambda_1}(\tau_1) \cdots X_n^{\lambda_n}(\tau_n)\right\}$$
$$= e^{(\alpha E)/T} \langle X_f^\alpha(\tau) X_1^{\lambda_1}(\tau_1) \cdots X_n^{\lambda_n}(\tau_n)\rangle_0$$
$$= e^{(\alpha E)/T} \sum_{i=1}^{n} (-1)^{p'_i} \langle X_1^{\lambda_1}(\tau_1) \cdots [X_f^\alpha(\tau), X_i^{\lambda_i}(\tau_i)]_\pm \cdots X_n^{\lambda_n}(\tau_n)\rangle_0$$
$$- e^{(\alpha E)/T} B_n.$$

Here p'_i is a number of the Fermi-type operators to the left of $X_i^{\lambda_i}(\tau_i)$ without counting the $X_f^\alpha(\tau)$ operator. The last expressions results in the formula

$$B_n = (1 - n_\alpha) \sum_{i=1}^{n} (-1)^{p'_i} \langle X_1^{\lambda_1}(\tau_1) \cdots [X_f^\alpha(\tau), X_i^{\lambda_i}(\tau_i)]_\pm \cdots X_n^{\lambda_n}(\tau_n) \rangle_0, \tag{5.15}$$

where

$$n_\alpha = [\exp(\alpha E)/T + 1]^{-1}. \tag{5.16}$$

Due to the even number of the Fermi-type operators in A_n it can be shown that

$$\pm (-1)^{p_n} B_n$$
$$= (1 - n_\alpha) \sum_{i=1}^{n} (-1)^{p_i} \langle X_1^{\lambda_1}(\tau_1) \cdots [X_f^\alpha(\tau), X_i^{\lambda_i}(\tau_i)]_\pm \cdots X_n^{\lambda_n}(\tau_n) \rangle_0.$$

In the interaction representation the time dependence is given by

$$[X_f^\alpha(\tau), X_i^{\lambda_i}(\tau_i)]_\pm = e^{\alpha E(\tau - \tau_i)} [X_f^\alpha, X_i^{\lambda_i}]_\pm^{\tau_i},$$

where the upper index τ_i means that an operator resulting from the calculation of the anticommutator (commutator) is written in the interaction representation with the time τ_i.

Finally the desired average (5.13) is equal to

$$A_n = \sum_{i=1}^{n} (-1)^{p_i} G_\alpha(\tau - \tau_i) \langle X_1^{\lambda_1}(\tau_1) \cdots [X_f^\alpha, X_i^{\lambda_i}]_\pm^{\tau_i} \cdots X_n^{\lambda_n}(\tau_n) \rangle_0, \tag{5.17}$$

with bare propagator

$$G_\alpha(\tau - \tau_i) = -\exp\{(\alpha E)(\tau - \tau_i)\} \times \begin{cases} n_\alpha & \tau > \tau_i \\ n_\alpha - 1 & \tau < \tau_i \end{cases}. \tag{5.18}$$

Equation (5.17) expresses the Wick's theorem that allows one to decrease the number of operators by one. Pairing of the operators X_f^α and $X_i^{\lambda_i}$ generates one new operator and the bare propagator. Due to the complicate algebra of X-operators this theorem differs from the similar one in the case of Fermi and Bose operators where the pairing of two operators gives c-number.

When the generating operator is of the Bose type the prove is similar and the Wick's theorem can be written as

$$-\langle T_\tau X_1^{\lambda_1}(\tau_1) X_2^{\lambda_2}(\tau_2) \cdots X_{k-1}^{\lambda_{k-1}}(\tau_{k-1}) X_f^\alpha(\tau) X_{k+1}^{\lambda_{k+1}}(\tau_{k+1}) \cdots X_n^{\lambda_n}(\tau_n) \rangle_0$$
$$= \sum_{i=1}^n G_\alpha(\tau - \tau_i) \langle T_\tau X_1^{\lambda_1}(\tau_1) \cdots [X_f^\alpha, X_i^{\lambda_i}]^{\tau_i} \cdots X_n^{\lambda_n}(\tau_n) \rangle_0.$$

Here the bare propagator describes a local Bose-type quasiparticle

$$G_\alpha(\tau - \tau_i) = \exp\{(\alpha E)(\tau - \tau_i)\} \times \begin{cases} n_\alpha & \tau > \tau_i \\ n_\alpha + 1 & \tau < \tau_i \end{cases}, \qquad (5.19)$$

with

$$n_\alpha = [\exp(\alpha E/T) - 1]^{-1}. \qquad (5.20)$$

To discuss a diagram formulation of the perturbation theory we introduce the notion of pairing of two operators. Pairing of the generating operator $X_f^\alpha(\tau)$ with arbitrary operator $X_g^\lambda(\tau')$ is shown by a line with arrow and results in a substitution of the two operators by their anticommutator (commutator), as given here

$$T_\tau \{ X_1^{\lambda_1}(\tau_1) \cdots \underbrace{X_f^\alpha(\tau) \cdots X_g^\lambda(\tau')}_{\longrightarrow} \cdots X_n^{\lambda_n}(\tau_n) \}$$
$$= (-1)^{P_\alpha} G_\alpha(\tau - \tau') \delta_{fg} T_\tau \{ X_1^{\lambda_1}(\tau_1) \cdots [X_g^\alpha, X_g^\lambda]_{\pm}^{\tau'} \cdots X_n^{\lambda_n}(\tau_n) \}.$$

Here P_α is the number of Fermi-type operator's permutations to get the operator X_f^α near and to the left of X_g^λ. With this notion we can formulate the Wick's theorem in the following way:

Wick's theorem: Taken with the sign "minus", the time ordered average of X-operators product with nondiagonal operators is equal to the sum of τ-ordered averages with subsequent pairings of the generating operator $X_f^\alpha(\tau)$ with all other operators.

For example, to calculate the average of τ-ordered product of four X operators we have to make the following pairing

$$-\langle T_\tau X_1^{\lambda_1}(\tau) X_2^{\lambda_2}(\tau_2) X_f^\alpha(\tau) X_4^{\lambda_4}(\tau_4) \rangle_0 = \langle T_\tau X_1^{\lambda_1}(\tau_1) X_2^{\lambda_2}(\tau_2) X_f^\alpha(\tau) X_4^{\lambda_4}(\tau_4) \rangle_0$$
$$+ \langle T_\tau X_1^{\lambda_1}(\tau_1) X_2^{\lambda_2}(\tau_2) X_f^\alpha(\tau) X_4^{\lambda_4}(\tau_4) \rangle_0 + \langle T_\tau X_1^{\lambda_1}(\tau_1) X_2^{\lambda_2}(\tau_2) X_f^\alpha(\tau) X_4^{\lambda_4}(\tau_4) \rangle_0.$$

In two last terms, the permutation of the generating operator $X_f^\alpha(\tau)$ with the pairing operators is counted in calculating P_α.

5.3 Averaging the Diagonal X-Operators

After several steps of applying the Wick's theorem, the calculated average is expressed by a sum of averages containing only the product of diagonal operators $h_{fn} = X_f^{nn}$. Thus the next problem is to calculate the average of the type

$$A_{f_1 f_2 \cdots f_m}^{n_1 n_2 \cdots n_m} = \langle X_{f_1}^{n_1 n_1} X_{f_2}^{n_2 n_2} \cdots X_{f_m}^{n_m n_m} \rangle_0.$$

The simplest average is

$$A_f^n = \langle X_f^{nn} \rangle_0 = Sp\{X_f^{nn} \exp(-H_0/T)\}/Sp\{\exp(-H_0/T)\}.$$

To calculate these averages we use the following trick: let us assume that all sites are nonequivalent and local energies become site-dependent. Then H_0 is written in the form

$$H_0 = \sum_f \sum_n E_{fn} X_f^{nn}$$

and the local operator average is given by the derivative

$$A_f^n = -T(\partial \ln Z_0/\partial E_{fn}),$$

where Z_0 is a statistical sum

$$Z_0 = Sp\{\exp(-H_0/T)\} = \prod_f \sum_n \exp(-E_{fn}/T).$$

A simple calculation gives

$$A_f^n = \exp(-E_{fn}/T)/\sum_m \exp(-E_{fm}/T). \qquad (5.21)$$

We denote $N_{fn} = \langle X_f^{nn} \rangle_0$ in the space uniform case $N_{fn} = N_n$. It will be displayed graphically as two concetric circles

$$N_n \longleftrightarrow \underset{n}{\textcircled{\circ}}.$$

The next step is the two diagonal operator's average

$$A_{fg}^{nm} = \langle X_f^{nn} X_g^{mm} \rangle_0.$$

It can be written in the following way

$$A_{fg}^{nm} = -(T)^2 \frac{1}{Z_0} \frac{\partial^2 Z_0}{\partial E_{fn} \partial E_{gm}} = (-T)\frac{1}{Z_0}\frac{\partial}{\partial E_{fn}}\left\{Z_0(-T)\frac{1}{Z_0}\frac{\partial Z_0}{\partial E_{gm}}\right\}$$

$$= (-T)\frac{1}{Z_0}\frac{\partial(Z_0 N_{gm})}{\partial E_{fn}} = N_{fn}N_{gm} + \delta_{fg}(\delta_{nm}N_{fn} - N_{fn}N_{gm}).$$

In the space uniform case, we have

$$\langle X_f^{nn} X_g^{mm} \rangle_0 = N_n N_m + \delta_{fg}(\delta_{nm}N_n - N_n N_m). \tag{5.22}$$

For the equivalent cells it is useful to introduce a local statistical sum

$$z_0 = \sum_n \exp(-E_n/T)$$

and a second order cumulant

$$\Gamma_2^{nm} \equiv -(T)^2 \frac{\partial^2 \ln z_0}{\partial E_n \partial E_m} = \delta_{nm} N_n - N_n N_m. \tag{5.23}$$

Then Eq. (5.22) may be written as follows

$$A_{fg}^{nm} = \langle X_f^{nn} X_g^{mm} \rangle_0 = N_n N_m + \delta_{fg} \Gamma_2^{nm}. \tag{5.24}$$

The X-operators algebra reveals itself in the expression (5.23)–(5.24). If $f \neq g$ then cumulant contribution is zero and $A_{fg}^{nm} = N_n N_m$ means two independent averages at different sites. If $f = g$ the two terms $N_n N_m$ in the right side of Eq. (5.23) are canceled and the result is $\delta_{mn} N_n$, that is non-zero only when $m = n$. Indeed by the X-operators algebra

$$\langle X_f^{nn} X_f^{mm} \rangle_0 = \langle \delta_{nm} X_f^{nn} \rangle_0 = \delta_{nm} N_n.$$

In term of diagrams, we will display graphically the cumulant $\delta_{fg}\Gamma_2^{nm}$ as the following oval

$$\delta_{fg}\Gamma_2^{nm} \longleftrightarrow \underset{fn \quad gm}{\boxed{\circ \quad \circ}}$$

and the average (5.22) can be written as

$$\langle X_f^{nn} X_g^{mm} \rangle_0 = \underset{n \quad m}{\circledcirc \; \circledcirc} + \underset{fn \quad gm}{\boxed{\circ \quad \circ}}. \tag{5.25}$$

Similar approach to the three-operator's average A^{nmp}_{fgl} results in

$$\langle X^{nn}_f X^{mm}_g X^{pp}_l \rangle_0 = \underset{n\ m\ p}{\odot\odot\odot} + \underset{n\ \ gm\ lp}{\odot\,(\circ\ \circ)} + \underset{m\ \ fn\ lp}{\odot\,(\circ\ \circ)}$$

$$+ \underset{p\ \ fn\ gm}{\odot\,(\circ\ \circ)} + \underset{fn\ gm\ lp}{(\circ\ \circ\ \circ)}, \qquad (5.26)$$

where the 3rd-order cumulant is given by

$$\Gamma^{nmp}_3 \equiv -(T)^3 \frac{\partial^3 \ln z_0}{\partial E_n \partial E_m \partial E_p}$$
$$= \delta_{np}(\delta_{mp} N_m - N_m N_p)$$
$$- N_n(\delta_{mp} N_p - N_p N_m) - N_p(\delta_{nm} N_m - N_n N_m),$$

$$\underset{fn\ gm\ lp}{(\circ\ \circ\ \circ)} \leftrightarrow \delta_{fg}\delta_{fl}\Gamma^{nmp}_3.$$

The analytical expression for the A^{nmp}_{fgl} is given by

$$A^{nmp}_{fgl} = N_n N_m N_p + \delta_{gl} N_n \Gamma^{mp}_2 + \delta_{fl} N_m \Gamma^{np}_2 + \delta_{fg} N_p \Gamma^{nm}_2 + \delta_{fg}\delta_{gl}\Gamma^{nmp}_3$$

We encounter in Eq. (5.26) with different ovals containing two and three small circles. In higher order cumulants one will encounter oval with more circles and also the contribution with several ovals. For example, the four-operator's term is displayed graphically as

$$\langle X^{nn}_f X^{mm}_g X^{pp}_l X^{qq}_r \rangle_0 = \underset{n\ m\ p\ q}{\odot\odot\odot\odot} + \underset{n\ m\ \ lp\ rq}{\odot\odot\,(\circ\ \circ)}$$

$$+ \underset{n\ p\ \ gm\ rq}{\odot\odot\,(\circ\ \circ)} + \underset{n\ q\ \ gm\ lp}{\odot\odot\,(\circ\ \circ)} + \underset{m\ p\ \ fn\ rq}{\odot\odot\,(\circ\ \circ)}$$

$$+ \underset{m\ q\ \ fn\ lp}{\odot\odot\,(\circ\ \circ)} + \underset{p\ q\ \ fn\ gm}{\odot\odot\,(\circ\ \circ)} + \underset{n\ \ gm\ lp\ rq}{\odot\,(\circ\ \circ\ \circ)}$$

$$+ \underset{m\ \ fn\ lp\ rq}{\odot\,(\circ\ \circ\ \circ)} + \underset{p\ \ fn\ gm\ rq}{\odot\,(\circ\ \circ\ \circ)} + \underset{q\ \ fn\ gm\ lp}{\odot\,(\circ\ \circ\ \circ)}$$

$$+ \underset{fn\ gm\ lp\ rq}{(\circ\ \circ\ \circ\ \circ)} + \underset{fn\ gm\ lp\ rq}{(\circ\ \circ)(\circ\ \circ)}$$

$$+ \underset{fn\ lp\ gm\ rq}{(\circ\ \circ)(\circ\ \circ)} + \underset{fn\ rq\ gm\ lp}{(\circ\ \circ)(\circ\ \circ)}.$$

After the 4-circle oval there are three terms here with two 2-circle ovals. In general, ovals with 2-, 3-, etc. circles may take place simultaneously.

Finally, we get the following diagrammatic series for arbitrary order average of the diagonal X-operators $A^{pq\ldots s}_{12\ldots n}$

$$\left\langle X_1^{pp} X_2^{qq} \cdots X_n^{ss} \right\rangle_0 = \bigcirc\!\bigcirc \cdots \bigcirc$$

$$+ \underbrace{(\circ\,\circ)\bigcirc\cdots\bigcirc + \bigcirc(\circ\,\circ)\cdots\bigcirc + \cdots + \bigcirc\cdots\bigcirc(\circ\,\circ)}_{((n-1)/2 \text{ 2-circle ovals and one 1-circle oval})}$$

$$+ \underbrace{(\circ\,\circ\,\circ)\cdots\bigcirc + \bigcirc(\circ\,\circ\,\circ)\cdots\bigcirc + \cdots + \bigcirc\cdots(\circ\,\circ\,\circ)}_{(n(n-1)(n-2)/6\text{-terms with only one oval of the third order})}$$

$$+ \cdots + \underbrace{(\circ\,\circ\,\cdots\,\circ)}_{\text{One oval of }n\text{-th order}} + \boxed{\text{terms with only two ovals}}$$

$$+ \boxed{\text{terms with only three ovals}} + \cdots + \boxed{\text{terms with only } M \text{ ovals}} \,. \quad (5.27)$$

Here rectangle containing only two ovals is a sum of all diagrams with two 2-circle ovals, two 3-circle ovals, one 2-circle and one 3-circle ovals etc. The similar meaning have rectangles with only three ovals etc., up to M ovals. Here M denotes the maximal possible number of ovals. For even n, $M = n/2$ and the last rectangle in Eq. (5.27) contains only $n/2$ 2-circle ovals. The number of such diagrams is equal to $(2M - 1)!!$.

For n odd there are sets of diagrams with maximal number of ovals. First set is given by diagrams with $(n-1)/2$ 2-circle ovals and one 1-circle oval, the number of such diagrams is equal to n!!. The second set is given by diagrams with $(n-3)/2$ 2-circle ovals and one 3-circle oval, the number of such diagrams is

$$(n-1)n!!/6.$$

A n-circle oval is given by a product of $(n-1)$ δ_{ij} factors and the n-th order cummulant

$$\underbrace{(\circ\,\circ\,\cdots\,\circ)}_{1p\,2q\,\cdots\,ns} \leftrightarrow \delta_{12}\delta_{23}\cdots\delta_{n-1,n}\Gamma_n^{pq\ldots s}, \quad \Gamma_n^{pq\ldots s}(-T)^n \frac{\partial \ln z_0}{\partial E_p \partial E_q \cdots \partial E_s}.$$

5.4 External and Internal Operators: Simplest Vertexes

Using Wick's theorem we can show graphically each term in the perturbation expansion (5.11). While a notion of pairing of the two operators $X_f^\alpha(\tau)$ and $X_{f'}^{\alpha'}(\tau)$ resulting in the propagator $G_\alpha(\tau - \tau')$ is similar to the case of Bose and Fermi operators, a generation of a new X-operator instead of the commutator (anticommutator) $\{X_f^\alpha, X_g^\lambda\}_\pm$ makes the situation more complicate in a way similar to the spin operator diagram technique (Vaks et al 1967a, 1967b; Izyumov et al 1974).

Let us introduce some new notions:

(1) We subdivide X-operators into external and internal. In the n-th order contribution to $D_{\alpha\beta}(f\tau; g\tau')$ given by Eq. (5.12), the operators $X_f^\alpha(\tau)$ and $X_g^{-\beta}(\tau')$ are external while all others X-operators related to H_{int} are internal. An entrance point with the coordinates (f, τ) and the root vector α in a graph stands for the operator $X_f^\alpha(\tau)$. An exit point (g, τ') denote the operator $X_g^{-\beta}(\tau')$ and is shown by a point with a thin crossed out line with an arrow as follows

Here arrow shows a direction of flow for the root vector β. The line is crossed out to show that it is not the propagator line. The propagator $G_\alpha(\tau - \tau')$ line is shown by a thin line between the entrance and exit points and is marked by the root vector. The arrow shows a direction to the exit point.

(2) The part of a diagram connected with a propagator line and (or) an interaction line is called a vertex. A number of legs, i.e. external propagators and interaction lines connected with a vertex, determines the type of vertex. There are several elementary vertices with two, three, and four legs (see below).

The bare Green's function $D_{\alpha\beta}^{(0)}(f\tau; g\tau')$ is determined by

$$= -\langle T_\tau X_f^\alpha(\tau) X_g^{-\beta}(\tau') \rangle_0.$$

The operator corresponding to the entrance point, here it is the $X_f^\alpha(\tau)$, is to be a generating operator for pairing. Here there is only one possible pairing, resulting in

$$-\langle T_\tau X_f^\alpha(\tau) X_g^{-\beta}(\tau')\rangle_0 = \langle T_\tau X_f^\alpha(\tau) X_g^{-\beta}(\tau')\rangle_0 = \delta_{fg} G_\alpha(\tau-\tau')\langle [X_g^\alpha, X_g^{-\beta}]_\pm^{\tau'}\rangle_0.$$

A result of commutations in the right part may be a new non-diagonal operator $X_f^{\alpha-\beta}$ or a diagonal X-operator. The non-diagonal operator is shown by additional line as follows

$$\begin{array}{c} \beta \\ \overset{\alpha}{\underset{f\tau}{\longrightarrow}} \overset{}{\underset{g\tau'}{\bigwedge}} \underset{}{\alpha\,\beta} \end{array}, \quad (5.28)$$

Here a three-leg non-diagonal vertex appears.

This example shows the conservation of the root vector in the vertex (Zaitsev 1975, 1976). The factor $N_{\alpha-\beta}$ given by Eq. (1.47) is ascribed to this vertex.

The diagonal operator appears only when $\alpha = \beta$ and the corresponding two-leg vertex is equal to $\delta_{\alpha\beta}(\boldsymbol{\alpha}, A_\alpha \boldsymbol{h})$ and shown by an empty circle

$$\begin{array}{c} \beta \\ \overset{\alpha}{\underset{f\tau}{\longrightarrow}} \underset{g\tau'}{\circ} \end{array}. \quad (5.29)$$

Up to now we have not considered the averaging procedure in calculating $D^{(0)}$. The result of averaging of any non-diagonal X-operator is equal to zero, so the graph (5.28) does not contribute to the bare Green's function. Averaging the vertex in Eq. (5.29) gives the expression $\delta_{\alpha\beta} b(\alpha)$, where the filling factor $b(\alpha)$ is equal to

$$b(\alpha) = \langle \boldsymbol{\alpha}, \hat{A}_{\bar{\alpha}} \boldsymbol{h}\rangle_0 = \sum_{ip}(\delta_{ip} - \delta_{iq})(A_{\bar{\alpha}})_{ij}\langle h_j\rangle_0 = \sum_i (\delta_{ip} - \delta_{iq})(\boldsymbol{A}_\alpha)_{ii}\langle h_i\rangle_0$$
$$= (A_\alpha)_{pp}\langle h_p\rangle_0 - (A_\alpha)_q\langle h_q\rangle_0 = N_p - (-1)^{\Delta_{pq}} N_q. \quad (5.30)$$

This definition is valid both for the Bose and the Fermi type Green's functions. In the specific case of the Bose type quasiparticle we shall use the next notation

$$b(\alpha) = B(\alpha) = N_p - N_q, \quad \alpha = \alpha(p,q), \qquad (5.31)$$

and in a case of the Fermi type quasiparticle, the filling factor is given by

$$F(\alpha) = N_p + N_q, \quad \alpha = \alpha(p,q). \qquad (5.32)$$

Finally, the zeroth-order H_{int} contribution to the Green's function is shown in diagram (5.29) and is equal to

$$D^{(0)}_{\alpha\beta}(f\tau, g\tau') = \delta_{fg} G_\alpha(\tau - \tau')\delta_{\alpha\beta} b(\alpha).$$

It is obvious also that the propagator line

$$\xrightarrow[f\tau \quad \quad g\tau']{\alpha}$$

corresponds to the expression

$$\delta_{fg} G_\alpha(\tau - \tau').$$

Fourier transforming the propagator $G_\alpha(\tau - \tau')$ as follows

$$G_\alpha(\tau - \tau') = T \sum_{\omega_n} \exp[-i\omega_n(\tau - \tau')] G_\alpha(\omega_n),$$

$$G_\alpha(\omega_n) = \frac{1}{2} \int_{-1/T}^{1/T} \exp[+i\omega_n(\tau)] G_\alpha(\tau) d\tau,$$

we obtain the non-zero component of $G_\alpha(\omega_n)$:

$$G_\alpha(\omega_n) = (i\omega_n + \alpha E)^{-1},$$

with $\omega_n = (2n+1)\pi T$ ($\omega_n = 2n\pi T$) for the Fermi (Bose) type Green's function.

By a Fourier transformation of δ_{fg} we can write the contribution of diagram (5.29) in the form

$$\frac{T}{N} \sum_{k,\omega_n} \exp\left\{ik(\mathbf{R}_f - \mathbf{R}_g) - i\omega_n(\tau - \tau')\right\} G_\alpha(\omega_n)\delta_{\alpha\beta} b(\alpha).$$

Finally, the zeroth-order contribution to the Fourier transform $D_{\alpha\beta}(\boldsymbol{k},\omega_n)$ is given by

$$D_{\alpha\beta}(\boldsymbol{k},\omega_n) = G_\alpha(\omega_n)\delta_{\alpha\beta}b(\alpha). \tag{5.33}$$

In a momentum representation the propagator line is denoted by $(\boldsymbol{k},\omega_n)$ and root vector α

$$\begin{array}{c}\alpha\\ \longrightarrow\\ \boldsymbol{k},\omega_n\end{array} \quad\Longleftrightarrow\quad G_\alpha(\omega_n)\ .$$

We call $G_\alpha(\omega_n)$ the bare propagator. The factor $\delta_{\alpha\beta}b(\alpha)$ in Eq. (5.33) is the results of the X-operator algebra.

A first order contribution to the Green's function is given by

$$\delta D^{(1)}_{\alpha\beta}(f\tau;g\tau') = -\frac{(-1)}{1!}\int_0^{1/T}\left\langle T_\tau X_f^\alpha(\tau)X_g^{-\beta}(\tau')H_{\text{int}}(\tau_1)\right\rangle_0 d\tau_1.$$

In the X-operator representation the interaction Hamiltonian has three components: transversal, transversal–longitudinal, and longitudinal. A transversal contribution has the form

$$\delta D^{(1),\perp}_{\alpha\beta}(f\tau;g\tau')$$
$$= \frac{1}{2}\sum_{11'}\sum_{\alpha_1\beta_1} V^{\alpha_1\beta_1}_{11'}\int_0^{1/T} d\tau_1 \left\langle T_\tau X_f^\alpha(\tau)X_g^{-\beta}(\tau')X_1^{\alpha_1}(\tau_1)X_{1'}^{\beta_1}(\tau_1)\right\rangle_0.$$

Here we can demonstrate the Wick's theorem. Starting with the generating operator $X_f^\alpha(\tau)$ (as a non-diagonal operator corresponding to the entrance point) we can make three pairings

$$\delta D^{(1),\perp}_{\alpha\beta}(f\tau;g\tau') = -\frac{1}{2}\sum_{11'}\sum_{\alpha_1\beta_1} V^{\alpha_1\beta_1}_{11'}$$

$$\times \int_0^{1/T} d\tau_1 \left\{\begin{array}{l}\langle T_\tau X_f^\alpha(\tau)X_g^{-\beta}(\tau')X_1^{\alpha_1}(\tau_1)X_{1'}^{\beta_1}(\tau_1)\rangle_0\\ +\langle T_\tau X_f^\alpha(\tau)X_g^{-\beta}(\tau')X_1^{\alpha_1}(\tau_1)X_{1'}^{\beta_1}(\tau_1)\rangle_0\\ +\langle T_\tau X_f^\alpha(\tau)X_g^{-\beta}(\tau')X_1^{\alpha_1}(\tau_1)X_{1'}^{\beta_1}(\tau_1)\rangle_0\end{array}\right\}.$$

$$\tag{5.34}$$

Operator of the transverse interaction

$$V^{\alpha_1\beta_1}_{11'}X_1^{\alpha_1}(\tau_1)X_{1'}^{\beta_1}(\tau_1)$$

will be shown by the interaction line

Operators structure in the transversal contribution can be shown as follows

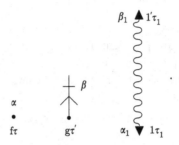

Different pairings with the generating operator $X_f^\alpha(\tau)$ after the first application of the Wick's theorem result in the following possible graphs:

$$\text{(a)} \quad \text{(b)} \quad \text{(c)} \quad \text{(d)} \tag{5.35}$$

Here the graphs (5.35(a), (b)) correspond to the first term in Eq. (5.34) for $\beta = \alpha$ and $\beta \neq \alpha$. Pairing of the generating operator $X_f^\alpha(\tau)$ with one of the internal X-operators gives the graph (5.35(c)) for

$$\alpha + \alpha_1 = 0 \quad \text{i.e.} \quad \alpha_1 = -\alpha,$$

if the result of commutation is a diagonal operator, and gives the graph (5.35(d)) when it is a non-diagonal operator $X^{\alpha+\alpha_1}$. The two graphs (c) and (d) contain two-leg and three-leg vertices

that look similar to the vertices (5.28) and (5.29), and they have the interaction line instead of the exit point as featured in (5.28, 5.29).

The last pairing in Eq. (5.34) results in the graphs that are topologically equivalent to (5.35(c)) and (5.35(d)), that gives a contribution to a multiplicative number of equivalent diagrams. It is worthwhile noting that for the Fermi type quasiparticles, the interaction matrix is obviously antisymmetric

$$V_{11'}^{\alpha_1 \beta_1} = -V_{1'1}^{\beta_1 \alpha_1}. \tag{5.36}$$

A change of sign when interchanging operators $X_{1'}^{\beta_1}(\tau_1)$ and $X_1^{\alpha_1}(\tau_1)$ in the topologically equivalent graph together with the antisymmetry condition (5.36) provides equal contribution of the equivalent graphs. For the Bose type quasiparticles, an equal contribution of two graphs with the inverted interaction line is evident.

An analytical expression for the graph (5.35(a)) is given by

$$\sum_{ij}(\alpha)_i (A_\alpha)_{ij} \langle T_\tau h_{gj}(\tau') X_1^{\alpha_1}(\tau_1) X_{1'}^{\beta_1}(\tau_1) \rangle_0. \tag{5.37}$$

Due to equivalency of both X-operators in the interaction line each of them can be chosen as a generating operator for the next pairing. Following the usual diagram technique for the Bose and the Fermi operators we take the second (typically annihilation) operator in H_{int} as the generating one.

Starting with the generating operator $X_{l'}^{\beta_1}(\tau_1)$, we can write Eq. (5.37) as

$$-\sum_{ij}(\alpha)_i(A_\alpha)_{ij}\langle T_\tau h_{gj}(\tau')X_1^{\alpha_1}(\tau_1)X_{1'}^{\beta_1}(\tau_1)\rangle_0$$
$$-\sum_{ij}(\alpha)_i(A_\alpha)_{ij}\langle T_\tau h_{gj}(\tau')X_1^{\alpha_1}(\tau_1)X_{1'}^{\beta_1}(\tau_1)\rangle_0. \quad (5.38)$$

The commutator of the non-diagonal operator $X_{l'}^{\beta_1}$ with the diagonal one h_{gj} is equal to $-(\beta_1)_j X_g^{\beta_1}$ and we can write the first term in Eq. (5.38) as follows

$$\mp \delta_{1'g} G_{\beta_1}(\tau_1 - \tau') \sum_{ij}(\alpha)_i(A_\alpha)_{ij}(\beta_1)_j \langle T_\tau X_g^{\beta_1}(\tau')X_1^{\alpha_1}(\tau_1)\rangle_0$$
$$= \mp \delta_{1'g} D_{\beta_1}(\tau_1 - \tau')(\alpha, A_\alpha \beta_1)\langle T_\tau X_g^{\beta_1}(\tau')X_1^{\alpha_1}(\tau_1)\rangle_0. \quad (5.39)$$

Here the upper sign corresponds to both Fermi type internal operators. To illustrate this term diagrammatically we start from (5.35(a)) and show the pairing by the propagator $G_{\beta_1}(1'\tau_1, g\tau')$. The final non-diagonal operator $X_g^{\beta_1}(\tau')$ is shown by a thin line with arrow and the root vector β_1:

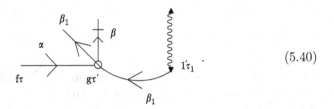

(5.40)

Here a new four-leg vertex appears with a numerical factor given by

$$\delta_{\alpha\beta}(\alpha, \hat{A}_\alpha \beta_1) \ . \quad (5.41)$$

We call this four-leg vertex "a pricked circle", and a generation of a four-leg vertex from a two-leg "empty circle" vertex is called the pricking of a circle.

5.5 A Hierarchy of Interaction Lines and a Topological Continuity Principles

To proceed with the calculation of expression (5.39) we use the Wick's theorem again. Both operators in Eq. (5.39) are non-diagonal and here we met a problem which operator should be generating. Difference choices of the generating operator result in different sets of diagrams. To avoid the ambiguity, Vedyaev and Nikolaev (1984) have proposed the principle of operator's hierarchy in their study of the Hubbard model. For two root vectors $(0,\sigma)$ and $(-\sigma, 2)$ of the Hubbard model this principle is quite effective (Izyumov et al 1994), but in the general case of arbitrary number of different quasiparticle excitations it is not convenient.

We propose another way to avoid the ambiguity in high-order diagrams. It includes a principle of interaction line's hierarchy (see below) and a **principle of topological continuity**. This principle claims:

(1) A non-diagonal X-operator appeared as a result of the k-th pairing is the generating operator in $(k+1)$-th pairing.
(2) When the result of the k-th pairing with the internal non-diagonal $X_{l'}^{\beta_1}$ operator from the transverse interaction line is a diagonal operator, the generating operator in the $(k+1)$-th pairing is the $X_l^{\alpha_1}$-operator connected to the diagram by the interaction line.
(3) When the result of the k-th pairing with the internal non-diagonal $X_l^{\alpha_1}$ operator from the transverse–longitudinal line is a diagonal operator, the generating operator in the $(k+1)$-th pairing is a non-diagonal operator from the other interaction line involved in the last previous pairing (see (5.46)).
(4) When conditions (1), (2) and (3) are not fulfilled the next generating operator is chosen by the principle of **interaction line hierarchy**: among several transverse, transverse–longitudinal and longitudinal lines pairing starts with the non-diagonal operators from the transverse line and only then from the transverse–longitudinal line (see (5.47)).

Here the transverse–longitudinal interaction line

$$\underset{1'_{\tau_1}}{\overset{n_1}{\circ}}\!\!\sim\!\!\sim\!\!\sim\!\!\sim\!\!\sim\!\!\sim\!\underset{1_{\tau_1}}{\overset{\alpha_1}{\blacktriangleright}} \qquad (5.42)$$

corresponds to the operators

$$V_{1'1}^{n_1\alpha_1} h_{1'n_1}(\tau_1) X_1^\alpha(\tau_1) \qquad (5.43)$$

and the longitudinal interaction line

$$\begin{array}{cc} n_1 & m_1 \\ \underset{1'\tau_1}{\circ\!\!\sim\!\!\sim\!\!\sim\!\!\sim\!\!\sim\!\!\sim\!\!\sim\!\!\circ} & \underset{1\tau_1}{} \end{array} \qquad (5.44)$$

is given by

$$V_{1'1}^{n_1 m_1} h_{1'n_1}(\tau_1) h_{1m_1}(\tau_1). \qquad (5.45)$$

Diagonal operator is denoted here as follows $X_f^{n,n} \equiv h_{f,n}$.

The meaning of the diagram of these two principles is quite simple. A propagator line is continuous up to a two-leg vertex (empty circle). If this vertex is originated from the transverse interaction line then the next propagator line continues from the opposite end of the interaction line. If the two-leg vertex is originated from the transverse–longitudinal line with a diagonal operator at the opposite end of this line, then the next pairing will start from a non-paired non-diagonal X-operator of the nearest interaction line involved in the previous pairing (point 3 in diagram (5.46))

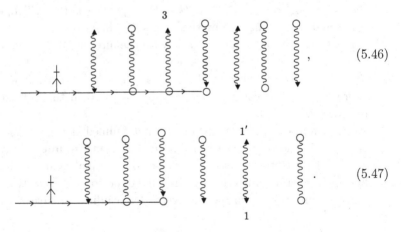

If the two-leg vertex is originated from the transverse–longitudinal line and there is no non-paired non-diagonal operator as shown in graph (5.47), then the next pairing starts from any of operators in the transverse line (1 or 1' in graph (5.47)). The two possible variants result in the topologically equivalent diagrams.

There is no operator hierarchy principle in our approach by the following reason. Each interaction line involves summation over root vectors of corresponding internal operators. Thus when we take one of the internal operators $X_n^{\alpha_n}(\tau_n)$ as a generating one the forthcoming summation over α_n involves all combinations of non-diagonal X-operators.

5.6 Calculation of the Sign for an Arbitrary Diagram

According to the topological continuity principle $X_g^{\beta_1}(\tau')$ is the generating operator in the next pairing for graph (5.40) — it is the pairing with the $X_l^{\alpha_1}(\tau_1)$ operator which is not zero only when $\beta_1 = -\alpha_1$. Thus the diagram for the first term of Eq. (5.38) is given by

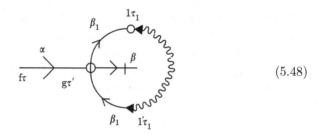
(5.48)

and is equal to

$$(\pm 1)\delta_{fg}\delta_{\alpha\beta}G_\alpha(\tau - \tau')\sum_{11'}\sum_{\alpha_1\beta_1}\left\{V_{11'}^{\alpha_1\beta_1}\delta_{1'g}\delta_{g1}\delta_{\beta_1,-\alpha_1}b(\beta_1)\right.$$
$$\left.\times \int_0^{1/T} d\tau_1 G_{\beta_1}(\tau_1 - \tau')G_{\beta_1}(\tau' - \tau_1)(\alpha\hat{A}_\alpha\beta_1)\right\}.$$

The sign of the diagram is determined by several factors. By the definition of the bare Green's function

$$D_{\alpha\beta}^{(0)}(f\tau, g\tau') = -\langle T_\tau X_f^\alpha(\tau)X_g^{-\beta}(\tau')\rangle_0 = \delta_{fg}G_\alpha(\tau - \tau')\delta_{\alpha\beta}b(\alpha),$$

it follows that each propagator line gives a factor of (-1). If G is the number of bare propagators then the diagram has the factor

$$(-1)^G.$$

A next factor is related to the four-leg vertex "pricked circles" that results from the commutation of the generating operator with a diagonal one,

$$\lfloor X_f^\alpha, X_g^{nn} \rfloor = -\alpha_n \delta_{fg} X_g^\alpha.$$

If A is the number of pricked circles then the corresponding factor is given by $(-1)^A$. A perturbation expansion of the scattering matrix in Eq. (5.11) gives a factor of $(-1)^{n+1}$ for an n-th-order diagram with n interaction lines.

Besides these factors there are four more complicated factors that require some new notions. Due to electron charge conservation law both root vectors in the transverse interaction $V_{fg}^{\alpha\beta}$ are of Fermi type or Bose type, we call the first interaction the line f-line and the second one the b-line. For the f-type interaction, the inversion of indices induces the change of sign (see Eq. (5.36)). So a question arises if this antisymmetry violates the additivity of the topologically equivalent diagrams. For example, two equivalent diagrams of the second order

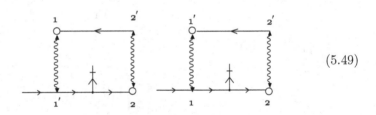

(5.49)

differs only in the inversion of the indices 1 and $1'$. For the left diagram we obtain

$$\pm \sum_{11'22'} \sum_{\alpha_1\beta_1\alpha_2\beta_2} \{ V_{11'}^{\alpha_1\beta_1} V_{22'}^{\alpha_2\beta_2} \delta_{f1'} \delta_{1'g} \delta_{g2} \delta_{2'1} N_{\alpha+\beta_1} N_{(\alpha+\beta_1)-\beta}$$

$$\times \delta_{-\alpha_2,(\alpha+\beta_1)-\beta} b(-\alpha_2) \delta_{-\alpha_1,\beta_2} b(\beta_2)$$

$$\times \int_0^{1/T} d\tau_1 \int_0^{1/T} d\tau_2 G_\alpha(\tau - \tau_1) G_{\alpha+\beta_1}(\tau_1 - \tau')$$

$$\times G_{(\alpha+\beta_1)-\beta}(\tau' - \tau_1) G_{\beta_2}(\tau_2 - \tau_1) \},$$

while the contribution of the right diagram is given by

$$\mp \sum_{11'22'} \sum_{\alpha_1\beta_1\alpha_2\beta_2} \{V_{11'}^{\alpha_1\beta_1} V_{22'}^{\alpha_2\beta_2} \delta_{f1}\delta_{1g}\delta_{g2}\delta_{2'1} N_{\alpha+\alpha_1} N_{(\alpha+\alpha_1)-\beta}$$

$$\times \delta_{-\alpha_2,(\alpha+\alpha_1)-\beta} b(-\alpha_2)\delta_{-\beta_1,\beta_2} b(\beta_2)$$

$$\times \int_0^{1/T} d\tau_1 \int_0^{1/T} d\tau_2 G_\alpha(\tau-\tau_1) G_{\alpha+\alpha_1}(\tau_1-\tau')$$

$$\times G_{(\alpha+\alpha_1)-\beta}(\tau'-\tau_1) G_{\beta_2}(\tau_2-\tau_1)\}.$$

There is a sign difference in the left and the right terms, because for the left diagram the generating operator $X_f^\alpha(\tau)$ has pairing with the second X-operator in H_{int} while for the right diagram the generating operator $X_f^\alpha(\tau)$ has pairing with the first X-operator in H_{int}. The permutation of two Fermi operators results in the additional factor (-1). If one substitute the summation indices as follow

$$1 \leftrightarrow 1', \qquad \alpha_1 \leftrightarrow \beta_1$$

then the difference is in the matrix $V_{11'}^{\alpha_1\beta_1}$ (left diagram) and $V_{1'1}^{\beta_1\alpha_1}$ (right diagram). Using Eq. (5.36), we find that they are equal. This example is a demonstration of the equal contribution of topological equivalent diagrams. On the other hand, this example reveals the sign correlation of diagram' and the order of indices in the interaction matrix elements. To calculate the corresponding sign factor for an arbitrary diagram we introduce a new definition of the interaction line connection:

(1) Connection of a transverse interaction line is a way of pairing of generating operator with one of the X-operators in $H_{\text{int}}^\perp(\tau_i)$ under that condition that $H_{\text{int}}^\perp(\tau_i)$ has not been involved in any previous pairing.
(2) Connection may be direct one when it occurs through the first X-operator in $H_{\text{int}}^\perp(\tau_i)$ and inverted one when it occurs through the second X-operators in $H_{\text{int}}^\perp(\tau_i)$. For example, in the left diagram (5.49) the first interaction line has an inverted connection while the second line has a direct connection. In the right diagram (5.49), both interaction lines have direct connections.

It turns out that each inverted connection results in one more permutation and gives an additional factor (-1). Let F_i is the total number of inverted connections of the transverse interaction lines of the Fermi type, then the corresponding sign factor is given by

$$(-1)^{F_i}.$$

The Fourier transform of Eq. (5.36) yields

$$V_q^{\alpha\beta} = -V_{-q}^{\beta\alpha}.$$

We will fix the position of matrix elements in the following way: all propagator lines and interaction lines in a diagram are noted by the wavenumber, Matsubara frequency and root vector conserving in each vertex. The position of matrix indices is determined by the direction of the wavevector propagation.

We fix the direction of the wavenumber in the following way

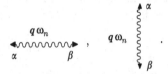

We define $V_q^{\alpha\beta}$ with vector q directed from left to right in the horizontal interaction line and from top to bottom in the vertical interaction line. The inverted connection takes place when the first connection of the interaction line to a propagator line is through the arrow of the wavenumber output.

It is evident that these notions depend on the direction of wavevector q but for a given diagram with fixed direction of wavevector these notions are unambiguous.

The next factor concerns diagrams of type (5.48), where after some steps involving the entrance and exit points the last pairing results in a diagonal operator and the next pairing should start from the isolated transverse interaction line $H_{\text{int}}^{\perp}(\tau_i)$. Here the second X-operator in $H_{\text{int}}^{\perp}(\tau_i)$ is chosen to be the generating one and it is permuted with the first X-operator, giving one more factor (-1). A closed loop of propagator and interaction lines that starts its formation with the transverse f-line, is called F-loop. Let K_F is the number of F-loops. Then the corresponding factor is equal to

$$(-1)^{K_F}.$$

For the diagram with more than one Fermi loop calculation of F_i should include all loops.

The next sign factor is determined by the crossing of the transverse interaction f-lines. The diagram with one crossing is given by the graph

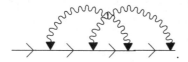.

Each crossing corresponds to the permutation of two Fermi-type X-operators and hence results in the factor (-1). Let P_F be the number of such crossings, then the corresponding factor is

$$(-1)^{P_F}.$$

The last sign factor arises in a situation when the exit operator permutes with some internal operators. In a diagram this situation means that along propagation from the entrance point to the exit point there are intersection with some arrows of interaction lines. If both the exit X-operator and the corresponding internal X-operator are of the Fermi-type then each intersection gives (-1). Let N_{eF} be the number of such intersections, then the factor $(-1)^{N_{eF}}$ arises. Finally, the sign of arbitrary diagram is equal to

$$(-1)^{G+A+n+1+F_i+K_F+P_F+N_{eF}}. \tag{5.50}$$

Thus for diagram (5.48) we have

$$G = 3, \quad A = 1, \quad n = 1, \quad F_i = 0, \quad P_F = 0, \quad N_{eF} = 0.$$

Here $K_F = 1$ for Fermi type interaction line and $K_F = 0$ for Bose type.

In the $(\boldsymbol{k}, \omega_n)$ representation, the contribution of the diagram (5.48) is given by

$$(-1)^{K_F} \delta_{\alpha\beta} G_\alpha(\omega_n) \frac{T}{N} \sum_{\boldsymbol{p}\omega_m \beta_1} V_{\boldsymbol{p}}^{-\beta_1,\beta_1} G_{\beta_1}^2(\omega_m)(\boldsymbol{\alpha}, \hat{A}_\alpha \boldsymbol{\beta}_1) b(\beta_1). \tag{5.51}$$

It is obvious that the graph (5.48) gives zero contribution due to the Kronecker deltas $\delta_{1'g} \delta_{g1}$ resulting in only single site contribution $V_{gg}^{-\beta_1,\beta_1} \equiv 0$. Indeed it is the main idea of the X-operator representation that all single site interactions are included in the local part of the Hamiltonian H_0 while the interatomic interactions and hopping are considered as perturbation. Nevertheless it is useful to study in detail the first order diagrams because they often are skeleton diagrams and give non-zero contribution after renormalization of the bare propagators, interaction lines and vertices in higher-order terms.

5.7 Diagrams with Ovals

As for the second term in Eq. (5.38), we have non-zero contribution only if $\beta_1 = -\alpha_1$. It is equal to

$$\delta_{\beta_1,-\alpha_1}\delta_{1'1}G_{\beta_1}(\tau_1 - \tau_1)\sum_{ijlm}(\alpha)_i(A_\alpha)_{ij}(\beta_1)_l(A_{\beta_1})_{lm}\langle h_{gj}h_{1m}\rangle_0.$$

Here an equal time propagator with

$$\tau_1 - \tau_1 = \lim(\tau_1 - \tau_1'), \qquad \tau_1' = \tau_1 + \delta, \qquad \delta \to +0$$

has the same definition as in the usual diagram technique for Fermi and Bose operators (Abrikosov et al 1965). Averaging a product of the two diagonal X-operators will result in

$$\delta_{\beta_1,-\alpha_1}\delta_{1'1}G_{\beta_1}(\tau_1 - \tau_1)[b(\alpha)b(\beta_1) + \delta_{g1}b(\alpha,\beta_1)], \qquad (5.52)$$

where a two-leg vertex with second-order cumulant Γ_2 is given by

$$b(\alpha,\beta_1) = \sum_{ijlm}(\alpha)_i(A_\alpha)_{ij}(\beta_1)_l(A_{\beta_1})_{lm}\Gamma_2^{jm} = (\boldsymbol{\alpha},\hat{A}_\alpha\hat{\Gamma}_2\hat{A}_{\beta_1}\boldsymbol{\beta}_1).$$

Here, the matrix $\hat{\Gamma}_2$ has the components (5.23). The two terms in Eq. (5.52) correspond to the next diagrams

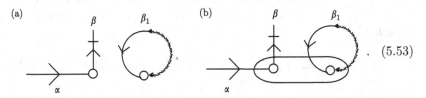

$$. \qquad (5.53)$$

The first graph here is a non-connected one with the usual definition of connected and non-connected diagrams: a non-connected diagram contains several parts which are not connected by any propagator, interaction line, and oval. The graph (5.53(b)) is the connected diagram, it arises from diagram (5.53(a)) by the unification of two small circle vertices in one oval vertex Γ_2. In high order diagrams these procedure can be generalized as follows: let p be the number of diagonal operators in the n-th order diagram. Then one of diagrams has p two-leg vertices (small circles). Other diagrams are obtained from the unification of two circles in one oval Γ_2, then three circles in oval Γ_3, etc., up to one oval Γ_p. Moreover, diagrams with two ovals Γ_2, with ovals Γ_2 and Γ_3, etc., will also appear. We call this procedure the oval growth.

The numerical value of both diagrams (5.53) is zero because in the first order diagram bare propagator is local while the interaction is nonlocal. Only high order renormalization resulting in the nonlocal propagators will give the non-zero contribution. Thus diagrams (5.53) and (5.48) are also skeleton. In the (k,ω) representation, diagram (5.53(a)) is given by

$$(-1)^{k_F}\frac{1}{2}\delta_{\alpha\beta}G_\alpha(\omega_n)b(\alpha)\sum_{\beta_1}\sum_{q\omega_m}V_q^{-\beta_1,\beta_1}G_{\beta_1}(\omega_m)b(\beta_1)e^{i\omega_m\delta}, \quad \delta\to+0.$$
(5.54)

A factor of $1/2$ appears in Eq. (5.54) because there are no two topological equivalent diagrams in this type of pairing. In general a numerical factor for a given diagram of n-th order can be calculated as

$$K = \frac{1}{n!}\left(\frac{1}{2}\right)^{N_1+N_2} N_{\text{top}}$$
(5.55)

where N_1 and N_2 are the numbers of the transverse and the longitudinal interaction lines respectively, and N_{top} is the number of topological equivalent diagrams. If we denote N_3 as the number of the transverse–longitudinal lines, then the order of diagram n is equal to

$$n = N_1 + N_2 + N_3.$$

The diagram (5.53(b)) is given by

$$(-1)^{k_F}\frac{1}{2}\delta_{\alpha\beta}G_\alpha(\omega_n)\frac{1}{N}\sum_{\beta_1}\sum_{q\omega_m}V_q^{-\beta_1,\beta_1}G_{\beta_1}(\omega_m)b(\alpha,\beta_1)e^{i\omega_m\delta}, \quad \delta\to+0.$$
(5.56)

Here the factor $1/N$ appears because inside the oval both lattice sites are equal and there is no additional sum over lattice site in comparison to

Eq. (5.54). Thus, both diagrams in graph (5.53) show the first order contribution to Green's function resulting from graph (5.35(a)).

Next pairing in the graph (5.35(b)) according to the principle of topological continuity gives only one possible diagram

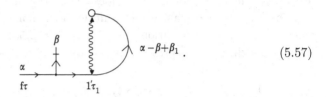
(5.57)

Here the pairing of the generating operator $X_g^{\alpha-\beta}(\tau')$ results in the non-diagonal operator $X^{(\alpha-\beta)+\beta_1}$. A vertex

$$\text{(5.58)}$$

that is similar to the vertex (5.28) is called non-diagonal three-leg vertex. Its numerical value is equal to $N_{\alpha+\beta}$. By the same reason as given above, diagram (5.57) is equal to zero and only high-order renormalization will give non-zero contribution. In the $(\mathbf{k}, \omega_n)$ representation this diagram can be written in the form

$$(-1)^{F+1} N_{\alpha-\beta} G_\alpha(\omega_n) G_{\alpha-\beta}(0) \frac{T}{N} \sum_{\beta_1} \sum_{\mathbf{q}\omega_m} V_\mathbf{q}^{-[(\alpha-\beta+\beta_1)],\beta_1} N_{(\alpha-\beta)+\beta_1}$$
$$\times G_{(\alpha-\beta)+\beta_1}(\omega_m) b[(\alpha-\beta)+\beta_1] e^{i\omega_m \delta}, \quad \delta \to +0$$

In the graph (5.35(c)) next pairing starts from the free end of the interaction line according to the principle of topological continuity. Three possible

diagrams are obtained

(a)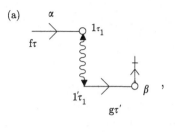

(b) 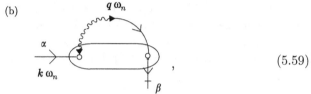   (5.59)

(c)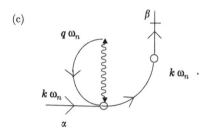

The analytical expression of diagram (5.59(a)) is

$$G_\alpha(\omega_n)b(\alpha)V_k^{-\alpha,\beta}G_\beta(\omega_n)b(\beta). \qquad (5.60)$$

When the two lattice sites, l and g, in (5.59(a)) are equal, $l = g$, it results in the oval diagram (5.59(b)), which is given by

$$G_\alpha(\omega_n)\frac{1}{N}\sum_q V_q^{-\alpha,\beta}G_\beta(\omega_n)b(\alpha,\beta).$$

Fourier transformation of the Kronecker symbol δ_{lg} gives an additional summation over q which may be interpreted as a wavevector (but not Matsubara frequency) "propagation" inside oval. For example, a quasiparticle $(\boldsymbol{k}, \omega, \alpha)$ due to the interaction $V_q^{-\alpha,\beta}$ create a new quasiparticle (q, ω, β). The remaining wavevector k–q "propagate" throw the oval. A contribution of the diagram (5.59(c)) is given by

$$G_\alpha(\omega_n)G_\beta(\omega_n)b(\beta)\frac{T}{N}\sum_{q,\omega_m} V_q^{-\alpha,\beta}G_\beta(\omega_m)(\alpha,\hat{A}_\alpha\beta).$$

Here we have $G = 3$, $A = 1$, $n = 1$, $P_F = 0$, $N_{eF} = 0$, $F_i = 0$ and positive sign of the diagram. The last class of first order diagrams results from (5.35(d)). According to topological continuity principle the generating operator is $X_1^{\alpha+\alpha_1}(\tau_1)$. Two different pairing are possible. At first, the propagation line $(\alpha + \alpha_1)$ starts from the internal three-leg vertex, comes to the exit point, forms an external three-leg vertex and stops at the free arrow of the interaction line (5.61(a)). Then the propagation line starts from the internal three-leg

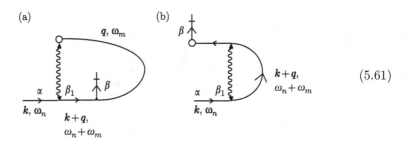

(5.61)

vertex and goes directly to the free arrow of the interaction line forming one more internal three-leg vertex and then stops in the exit point (5.61(b)). Both diagrams are of skeleton type which equal to zero in the first order of perturbation theory. (Here $n = 1$, $G = 3$, $A = 0$.) If both β and β_1 are of Fermi type then $N_{eF} = 1$, otherwise $N_{eF} = 0$ for diagram (5.61(a)). For diagram (5.61(b)), we have $N_{eF} = 0$ or $N_{eF} = 2$ depending on the type of quasiparticles, hence in each case $(-1)^{N_{eF}} = 1$. The corresponding analytical expressions are given by

$$(-1)^{F_i+P_e+1}G_\alpha(\omega_n)$$
$$\times \frac{T}{N}\sum_{q,\omega_m,\beta_1,\alpha_1} V_q^{\alpha_1,\beta_1} N_{\alpha+\beta_1} G_{\alpha+\beta_1}(\omega_n+\omega_m)$$
$$\times N_{(\alpha+\beta_1)-\beta}G_{(\alpha+\beta_1)-\beta}(\omega_m)b(-\alpha_1)\delta(\alpha_1+((\alpha+\beta_1)-\beta))$$

and

$$(-1)^{F_i+1} G_\alpha(\omega_n) G_\beta(\omega_n) b(\beta)$$
$$\times \frac{T}{N} \sum_{\mathbf{q},\omega_m,\beta_1,\alpha_1} V_\mathbf{q}^{\alpha_1,\beta_1} N_{\alpha+\beta_1} G_{\alpha+\beta_1}(\omega_n + \omega_m) N_{(\alpha+\beta_1)-\alpha_1}$$
$$\times \delta[((\alpha+\beta_1) - \alpha_1) - \beta].$$

Thus we have considered all first order diagrams steming from the transverse interaction. Diagrams resulted from longitudinal interaction are described below.

The first order contribution to Green's function induced by the longitudinal interaction can be written as

$$-\left(\frac{-1}{1!2}\right) \sum_{11'n_1m_1} V_{11'}^{n_1m_1} \int_0^{1/T} d\tau_1 \langle T_\tau X_f^\alpha(\tau) X_g^{-\beta}(\tau') X_1^{n_1}(\tau_1) X_{1'}^{m_1}(\tau_1) \rangle_0.$$

(5.62)

The initial operator configuration in this case looks like

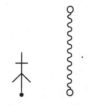

Application of the Wick's theorem results in the following diagrams (5.63):

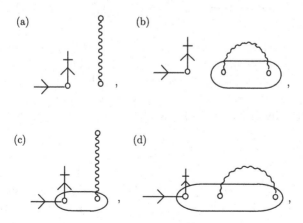

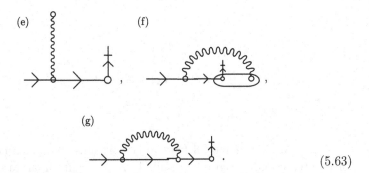

(5.63)

Here the first three diagrams are obtained from the initial operator configuration through the formation of propagation line from the entrance point to the exit point due to pairing of the generating operator $X_f^\alpha(\tau)$ with the operator $X_g^{-\beta}(\tau')$. As a result of this procedure, the average of the product of three diagonal operators depends on the 3 lattice sites g, 1, and $1'$. The diagram (5.63(a)) corresponds to the first term (5.26). Oval growth procedure results in the diagrams (b), (c), and (d) in Eq. (5.63). These diagrams can be calculated with the help of Eqs. (5.25). The analytical expressions for them may be written as

for (5.63a): $\quad -\dfrac{1}{2}\delta_{\alpha\beta}G_\alpha(\omega_n)b(\alpha)\left(\dfrac{N}{T}\right)\displaystyle\sum_{n_1 m_1} V_0^{n_1 m_1} N_{n_1} N_{m_1},$

for (5.63b): $\quad -\delta_{\alpha\beta}G_\alpha(\omega_n)\left(\dfrac{1}{T}\right)\displaystyle\sum_{n,m,n_1,m_1} V_0^{n_1 m_1}(\alpha)_n(\hat{A}_\alpha)_{nm}\Gamma_2^{nm_1} N_{m_1},$

for
(5.63c): $\quad -\dfrac{1}{2}\delta_{\alpha\beta}G_\alpha(\omega_n)\left(\dfrac{1}{NT}\right)\displaystyle\sum_{q}\sum_{n,m,n_1,m_1} V_q^{n_1 m_1}(\alpha)_n(\hat{A}_\alpha)_{nm}\Gamma_3^{mn_1 m_1}.$

The diagram (5.63(e)) results from pairing of the generating operator $X_f^\alpha(\tau)$ with one of the diagonal X-operator from H_{int} and the next pairing of thus appeared new non-diagonal X-operator with the exit operator $X_g^{-\beta}(\tau')$. Here we use the principle of topological continuity formulated above. Calculation of the diagram (5.63(e)) gives

$$-\delta_{\alpha\beta}G_\alpha^2(\omega_n)b(\alpha)\sum_{n_1 m_1} V_0^{n_1 m_1}(\alpha)_{n_1} N_{m_1}.$$

Here the factor $(\alpha)_{n_1}$ corresponds to the three-leg pricked circle vertex. Diagram (5.63(f)) is a consequence of diagram (5.63(e)) due to the oval

growth procedure. It is equal to

$$-\delta_{\alpha\beta}G_\alpha^2(\omega_n)\left(\frac{1}{N}\right)\sum_{\boldsymbol{q}}\sum_{n,m,n_1,m_1} V_{\boldsymbol{q}}^{n_1 m_1}(\boldsymbol{\alpha})_{n_1}(\boldsymbol{\alpha})_n(\hat{A}_{\boldsymbol{\alpha}})_{nm}\Gamma_2^{n\,m_1}.$$

The last diagram in Eq. (5.63) is given by the combination of two diagonal three-leg vertices. A straightforward calculation yields

$$-\delta_{\alpha\beta}G_\alpha^2(\omega_n)b(\alpha)\left(\frac{T}{N}\right)\sum_{\omega_m\boldsymbol{q}}\sum_{n_1,m_1} V_{\boldsymbol{q}}^{n_1 m_1}(\boldsymbol{\alpha})_{n_1}(\boldsymbol{\alpha})_{m_1}G_\alpha(\omega_m)e^{i\omega_m\delta},$$

$\delta \to +0$.

In some models the transverse–longitudinal interaction occurs, for example $V_{ij}S_i^+ S_j^z$ in the anisotropic Heisenberg model. The corresponding initial operators configuration looks like

All possible first order diagrams are given by Eq. (5.64). To obtain these diagrams we start pairing the generating entrance operator $X_f^\alpha(\tau)$. First line in graph (5.64) shows diagrams where pairing takes place with the exit operator $X_g^{-\beta}(\tau')$. Here the non-diagonal three-leg vertex arises, otherwise one remaining non-diagonal non-paired X-operator will give zero after averaging. The second pairing can occur either with non-diagonal X-operator (diagrams 5.64(a) and (b)) or with diagonal X-operator (diagram 5.64(c)).

(5.64)

Next five diagrams in graph (5.64) arise when the entrance operator $X_f^\alpha(\tau)$ pairs with the non-diagonal (5.64(d),(e),(f)) or the diagonal (5.64(g),(h)) X-operators from the interaction line. In the first case only non-diagonal three-leg vertex gives non-zero contribution. The second pairing with the exit X-operator results in diagram (5.64(d)) and in diagram (5.64(e)) after the oval growth procedure, while the second pairing with the diagonal X-operator of the interaction line gives diagram (5.64(f)). In the second case, pairing of X_f^α with diagonal operator results in a new non-diagonal X-operator which is the generating one in the second pairing either with the exit operator (5.64(g)) or with the non-diagonal X-operator of the interaction line (5.64(h)).

It should be mentioned that both the longitudinal and transverse–longitudinal interaction lines are Bose type lines, because any diagonal X-operator is of Bose type. In the simple models of SEC such as the Hubbard model there are only transverse Fermi and Bose interaction lines. Bose type interaction lines can arise for example in generalized Hubbard model taking into account interatomic Coulomb interaction $V_{ij}n_i n_j$ contains both Fermi type interaction (interatomic hopping) and Bose interaction (interatomic Coulomb) lines. The transverse–longitudinal interaction lines take place for the antiferromagnets in the magnetic field and for the anisotropic ferromagnets, when non-colinear magnetic structure is generated.

5.8 General Rules for Arbitrary Order Diagram

The description of first order diagrams given in detail above can be generalized for arbitrary order diagrams. The following rules can be formulated to obtain the n-th order contribution to the Green's function $D_{\alpha\beta}(\boldsymbol{k}, \omega_n)$:

1. To draw all initial operator configurations with entrance and exit points and n interaction lines. If there are m types of interaction lines ($m \leq 3$) then the number of different initial configurations is equal to

$$\frac{(n+m-1)!}{n!(m-1)!}.$$

2. Using the principles of topological continuity and hierarchy of interaction lines to display graphically all connected diagrams. A propagator line starts at the entrance point and is connected to interaction lines and the exit point. New propagator lines are generated up to the

situation when all arrows of interaction lines and the exit point are connected in one graph. It means the absence of non-paired non-diagonal X-operators.

3. Each connection of a propagator line to interaction line and exit point forms a vertex. Propagation line connected to the arrow of interaction line or to the exit point can either stop forming a two-leg vertex or continue forming a non-diagonal three-leg vertex. Propagation line connected to the circle (diagonal end) of interaction line forms a diagonal three-leg vertex. Propagation line can prick a circle in two-leg vertex transforming it into a four-leg pricked circle vertex. For a given set of connections different diagrams are obtained by a procedure of oval growth.

4. Each propagator and interaction line is labelled by a value of quasi-momentum, Matsubara frequency and root vector. All these parameters are conserved in each vertex. The start and exit propagator lines are labelled by parameters k, ω_n, α and k, ω_n, β, correspondingly.

5. A propagator line with parameters k, ω_n, γ is given by

$$G_\gamma(\omega_n) = \frac{1}{i\omega_n + \gamma E}.$$

While bare propagator is local and does not depend on momentum the forthcoming renormalization results in a k-dependence of the dressed Green's function.

6. Interaction lines are given by the following matrix elements: $V_q^{\alpha\beta}$ (transverse), V_q^{nm} (longitudinal), $V_q^{\alpha m}$ (transverse-longitudinal). The direction of momentum q propagation indicates the order of root vectors for the transverse interaction line. For example, in the following graph

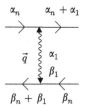

the interaction line is given by $V_q^{\alpha_1\beta_1}$ if the vector q is directed downwards and by $V_q^{\beta_1\alpha_1}$ if q directed upwards. Conventional directions of q in $V_q^{\alpha_1\beta_1}$ are from left to right and from top to bottom.

7. The first few vertices are

   (i) two-leg vertex

   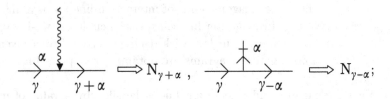

   (ii) nondiagonal three-leg vertex

   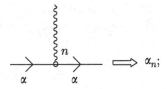

   (iii) diagonal three-leg pricked circle vertex

   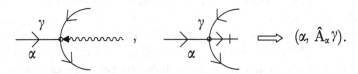

   (iv) four-leg pricked circle vertex

   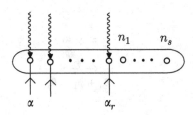

8. Oval with r two-leg vertices and s diagonal operators (small circles) is given by

   is given by

   $$(\alpha_1)_{i_1}(A_{\alpha_1})_{i_1 j_1}(\alpha_2)_{i_2}(A_{\alpha_2})_{i_2 j_2}\cdots(\alpha_r)_{i_r}(A_{\alpha_r})_{i_r j_r} \times \Gamma_{r+s}^{j_1 j_2 \cdots j_r n_1 n_2 \cdots n_s}.$$

9. The integration over all internal quasi-momenta and summation over all internal Matsubara frequencies and root vectors are carried out. A numerical coefficient for a diagram can be written as

$$(-1)^Z \left(\frac{T}{N}\right)^{G-n-1} \left(\frac{1}{N}\right)^{N_\Gamma} \frac{P_{n_1 n_2 \cdots}}{n_1! \, n_2! \cdots},$$

with the sign factor Z given by

$$Z = G + A + n + 1 + F_i + K_F + P + N_{eF}.$$

Here, N_Γ is determined by

$$N_\Gamma = N(\Gamma_2) + 2N(\Gamma_3) + 3N(\Gamma_4) + \cdots,$$

where $N(\Gamma_i)$ is the number of ovals with i circles and two-leg vertices. The number of interaction lines of the i-type is denoted n_i; $P_{n_1 n_2 \cdots}$ is a number of permutation of the same type of interaction lines, resulting in topological equivalent but different systems of pairing and ovals. For example in the following 4-th order diagram

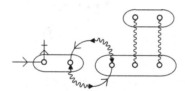

the number of the transverse (longitudinal) interaction lines is $n_1 = 2$ ($n_2 = 2$). Permutation of the transverse lines results in topologically equivalent but different diagrams. Contrary, permutation of the longitudinal lines results in the same diagram. Thus the coefficient

$$\frac{P_{n_1 n_2}}{n_1! n_2!}$$

for this diagram is equal to $2!/2!2! = 1/2$.

5.9 The Larkin Equation

In this section a general structure of the Green's function $D_{\alpha\beta}(\mathbf{k}, \omega_n)$ is analyzed. We consider separately two following cases:

(1) root vectors α and β are of the Fermi type;
(2) root vectors α and β are of the Bose type.

Let us consider the following diagrams for the Fermi type case.

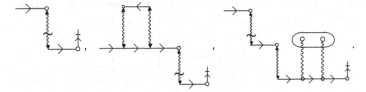

It is evident that each diagram above contains at least one transverse interaction line in such manner that the cutting of this line (shown above by a short wavy line) breaks the diagram resulting in two disconnected parts. None of these parts has both the entrance and exit points. Diagrams of this type are called reducible by Larkin. A diagram is irreducible if cutting across the transverse interaction line does not result in two disconnected part each containing the entrance or exit points. The last condition means that the next diagram

is an irreducible one in spite of the possibility to cut it through the transverse interaction line. If we cut this diagram then a first fragment

contains both the entrance and exit points. We note also that the remaining second part

can be connected to the first one only by one interaction line. Such diagrams are called one-leg diagrams and describe internal self-consistent fields. Summation of all one-leg diagrams results in the renormalization of the intra-cell parameters in H_0. Similar renormalizations in the case of longitudinal one-leg diagrams for spin systems are presented in the book by Izyumov, Kassan-Ogly and Skryabin (1974).

It is evident that each diagram for $D_{\alpha\beta}(\boldsymbol{k},\omega_n)$ is either irreducible or contains several irreducible parts connected by transverse interaction line. We call a sum of all irreducible diagrams the irreducible part of the Green's function $D_{\alpha\beta}(\boldsymbol{k},\omega_n)$ and show it as a large circle $T^{\alpha\beta}$.

$$\left(T_{\alpha\beta}\right) \Longrightarrow T_{\alpha\beta}(\boldsymbol{k},\omega_n)$$

Analytical formulae for diagrams from $T^{\alpha\beta}$ are the same as that for the Green's function with minor difference in sign. It is more convenient to determine a sign of a diagram from $T^{\alpha\beta}$ in a following way

$$(-1)^{Z'}, \quad Z' = Z - 1$$

with the value Z is given by Eq. (5.133). With the help of the irreducible parts, the total set of diagrams for the Green's function looks like

$$\begin{aligned}\xrightarrow[\alpha]{}\xrightarrow[\beta]{} &= \left(T^{\alpha\beta}\right) + \left(T^{\alpha\alpha_1}\right)\sim\sim\sim\left(T^{\beta_1\beta}\right) \\ &+ \left(T^{\alpha\alpha_1}\right)\sim\sim\sim\left(T^{\beta_1\alpha_2}\right)\sim\sim\sim\left(T^{\beta_2\beta}\right) + \cdots \\ &= \left(T^{\alpha\beta}\right) + \left(T^{\alpha\alpha_1}\right)\sim\sim\sim\xrightarrow[\beta_1]{}\xrightarrow[\beta]{}\, .\end{aligned} \quad (5.65)$$

Analytical expression for these diagrams yields

$$\begin{aligned}D_{\alpha\beta}&(\boldsymbol{k},\omega_n) \\ &= -T_{\alpha\beta}(\boldsymbol{k},\omega_n) + \sum_{\alpha_1\beta_1} T_{\alpha\alpha_1}(\boldsymbol{k},\omega_n)V_{\boldsymbol{k}}^{\alpha_1\beta_1}T_{\beta_1\beta}(\boldsymbol{k},\omega_n) \\ &\quad - \sum_{\alpha_1\beta_1\alpha_2\beta_2} T_{\alpha\alpha_1}(\boldsymbol{k},\omega_n)V_{\boldsymbol{k}}^{\alpha_1\beta_1}T_{\beta_1\alpha_2}(\boldsymbol{k},\omega_n)V_{\boldsymbol{k}}^{\alpha_2\beta_2}T_{\beta_2\beta}(\boldsymbol{k},\omega_n) + \cdots \\ &= -T_{\alpha\beta}(\boldsymbol{k},\omega_n) - \sum_{\alpha_1\beta_1} T_{\alpha\alpha_1}(\boldsymbol{k},\omega_n)V_{\boldsymbol{k}}^{\alpha_1\beta_1}D_{\beta_1\beta}(\boldsymbol{k},\omega_n).\end{aligned} \quad (5.66)$$

In matrix notations, Eq. (5.66) reads

$$\hat{D}(\boldsymbol{k},\omega_n) = -\hat{T}(\boldsymbol{k},\omega_n) - \hat{T}(\boldsymbol{k},\omega_n)\hat{V}_{\boldsymbol{k}}\hat{D}(\boldsymbol{k},\omega_n) \quad (5.67)$$

and its solution is given by

$$\hat{D}(\boldsymbol{k},\omega_n) = -\left[1+\hat{T}(\boldsymbol{k},\omega_n)\hat{V}_{\boldsymbol{k}}\right]^{-1}\hat{T}(\boldsymbol{k},\omega_n) = -\left[\hat{T}^{-1}(\boldsymbol{k},\omega_n)+\hat{V}_{\boldsymbol{k}}\right]^{-1}. \tag{5.68}$$

Equation (5.68) is called the Larkin equation. It transforms the calculation of Green's function G to the calculation of its irreducible part T. Practically one has to find an inverse matrix solving the Larkin equation. It is not a problem for systems with small number of intracell eigenstates like the Hubbard model. But for a system with large number N of eigenstates the corresponding dimension N^2 grow rapidly.

We can simplify this problem by analyzing the structure of interaction matrix V. Let us consider the intercell electron hopping Hamiltonian

$$\sum_{fg}\sum_{ij} t^{ij}_{fg} a^+_{fi} a_{gj} \tag{5.69}$$

where indices i and j denote all intracell quantum numbers including spin σ. Using the X-representation for electron operator

$$a_{fi} = \sum_{\alpha} \gamma_i(\alpha) X^{\alpha}_f,$$

we write Eq. (5.69) in the next form

$$\frac{1}{2}\sum_{fg}\sum_{ij}\sum_{\alpha\beta}\{t^{ij}_{fg}\gamma^*_i(-\alpha)\gamma_j(\beta) - t^{ji*}_{fg}\gamma^*_i(-\beta)\gamma_j(\alpha)\}X^{\alpha}_f X^{\beta}_g. \tag{5.70}$$

Here we use the Hermitian property of the Hamiltonian (5.69), that results in

$$t^{ji}_{gf} = t^{ij*}_{fg}.$$

We introduce the vector $c(\alpha)$ with components

$$c(\alpha) = [\gamma_1(\alpha), \gamma_2(\alpha), \ldots, \gamma^*_1(-\alpha), \gamma^*_2(-\alpha), \ldots], \tag{5.71}$$

a hopping matrix $\hat{t}_{fg}$,

$$(\hat{t}_{fg})_{ij} = t^{ij}_{fg}$$

and one more matrix

$$\hat{U}_{fg} = \begin{pmatrix} \hat{t}_{fg} & 0 \\ 0 & -\hat{t}^*_{fg} \end{pmatrix}.$$

We also define the scalar product of c-vectors as

$$\{c(\alpha), c(\beta)\} \equiv \sum_i c_i^*(\alpha) c_i(\beta).$$

The hopping Hamiltonian (5.70) now looks like

$$\frac{1}{2} \sum_{fg} \sum_{\alpha\beta} \{c(-\alpha), \hat{U}_{fg} c(\beta)\} X_f^\alpha X_g^\beta. \tag{5.72}$$

In this notation the intercell interaction matrix

$$V_{fg}^{\alpha\beta} = \{c(-\alpha), \hat{U}_{fg} c(\beta)\} \tag{5.73}$$

is written as a product of terms depending separately on α and β. This split character of $V^{\alpha\beta}$ allows one to write $D_{\alpha\beta}$ with the help of $T_{\alpha\beta}$ and a vector Z_β

$$Z_\beta(\mathbf{k}, \omega_n) = \sum_{\beta_1} c(\beta_1) D_{\beta_1\beta}(\mathbf{k}, \omega_n).$$

The Larkin equation now reads

$$D_{\alpha\beta}(\mathbf{k}, \omega_n) = -T_{\alpha\beta}(\mathbf{k}, \omega_n) - \sum_{\alpha_1} T_{\alpha\alpha_1}(\mathbf{k}, \omega_n) \{c(-\alpha_1), \hat{U}_\mathbf{k} Z_\beta(\mathbf{k}, \omega_n)\}.$$

Multiplying each term by $c(\alpha)$ and summing over α we obtain the equation for vector Z:

$$Z_\beta(\mathbf{k}, \omega_n) = -\sum_\alpha c(\alpha) T_{\alpha\beta}(\mathbf{k}, \omega_n)$$
$$- \sum_{\alpha\alpha_1} c(\alpha) T_{\alpha\alpha_1}(\mathbf{k}, \omega_n) \{c(-\alpha_1), \hat{U}_\mathbf{k} Z_\beta(\mathbf{k}, \omega_n)\}.$$

The i-th component of Z_β looks like

$$(Z_\beta)_i = -\sum_\alpha c_i(\alpha) T_{\alpha\beta} - \sum_{\alpha\alpha_1 jl} c_i(\alpha) T_{\alpha\alpha_1} c_j(-\alpha_1), (\hat{U}_\mathbf{k})^{jl} (Z_\beta)_l.$$

To solve this equation we introduce a matrix

$$L_{ij}(\mathbf{k}, \omega_n) = \sum_{\alpha\alpha_1} c_i(\alpha) T_{\alpha\alpha_1}(\mathbf{k}, \omega_n) c_j(-\alpha_1).$$

With the matrix L, the equation for Z matrix can be written as

$$\{1 + \hat{L}(\mathbf{k}, \omega_n) \hat{U}_\mathbf{k}\} Z_\beta(\mathbf{k}, \omega_n) = -\sum_\alpha c_i(\alpha) T_{\alpha\beta}(\mathbf{k}, \omega_n).$$

Its solution is given:

$$Z_\beta = \{1 + \hat{L}\hat{U}_k\}^{-1} B_\beta,$$

where the vector B is equal to

$$B_\beta = -\sum_\alpha c(\alpha) T_{\alpha\beta}.$$

Finally we write the Larkin equation in the form

$$\begin{aligned}D_{\alpha\beta}(\boldsymbol{k},\omega_n) &= -T_{\alpha\beta}(\boldsymbol{k},\omega_n) \\ &- \sum_{\alpha_1} T_{\alpha\alpha_1}(\boldsymbol{k},\omega_n)\{c(-\alpha_1), [\hat{U}_k^{-1} + L]^{-1} B_\beta\}\end{aligned} \quad (5.74)$$

where the matrix dimension is determined by the number of different parameters i (band indices) in the X-operator representation. This number is usually not large in comparison with the number of different root vectors α. Thus, the presentation of the Larkin equation in the form (5.74) results from the split form of the interaction matrix.

Now we consider the Green's function $D_{\alpha\beta}$ with Bose-type vectors α and β. Besides diagrams that are similar to that of the Fermi-type there are some new type of diagrams provided by the longitudinal and transverse–longitudinal interaction lines. Some of them are shown below

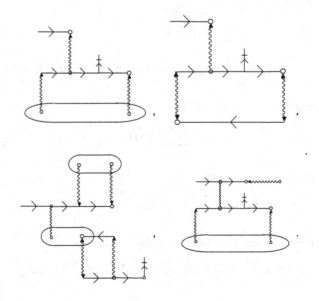

It is clear that it is possible to cut some diagrams through the longitudinal and transverse–longitudinal interaction lines as well as through the transverse lines. We have to introduce new irreducible parts $T^{\gamma n}, T^{n\gamma}, T^{nm}$. In this case it is convenient to use the matrix notation

$$D_{\lambda\lambda'}(f\tau; g\tau') = -\langle T_\tau \tilde{X}_f^\lambda(\tau)\tilde{X}_g^{-\lambda'}(\tau')\rangle,$$

where λ can be a root vector α or an index of the intracell state n. If $\lambda = n$ then

$$X_g^\lambda = X_g^{-\lambda} = X_g^{nn}$$

A mixed Green's function $\hat{D}_{\alpha n}$ with a Fermi-type vector α is equal to zero by definition.

We introduce four irreducible parts for the Bose-type Green's function

$\left(T^{\alpha\beta}\right)$, $\left(T^{\alpha m}\right)$, $\left(T^{n\beta}\right)$, $\left(T^{nm}\right)$,

in a matrix notation as shown by

$\left(T^{\lambda\lambda'}\right)$,

The Larkin equation looks like

(5.75)

When an index λ is given by the root vector α then we should draw an arrow at the corresponding end of the interaction line. We introduce the following block matrices for $\hat{T}(\boldsymbol{k}, \omega_n)$

$$\hat{T}(\boldsymbol{k},\omega_n) = \begin{pmatrix} T^{\alpha\beta}(\boldsymbol{k},\omega_n) & T^{\alpha l}(\boldsymbol{k},\omega_n) \\ T^{m\beta}(\boldsymbol{k},\omega_n) & T^{ml}(\boldsymbol{k},\omega_n) \end{pmatrix} \quad (5.76)$$

for the Green's function

$$\hat{D}(\boldsymbol{k},\omega_n) = \begin{pmatrix} D_{\alpha\beta}(\boldsymbol{k},\omega_n) & D_{\alpha l}(\boldsymbol{k},\omega_n) \\ D_{m\beta}(\boldsymbol{k},\omega_n) & D_{ml}(\boldsymbol{k},\omega_n) \end{pmatrix} \quad (5.77)$$

and for the interaction

$$\hat{V}_k = \begin{pmatrix} V_k^{\alpha\beta} & V_k^{\alpha l} \\ V_k^{m\beta} & V_k^{ml} \end{pmatrix}. \tag{5.78}$$

Using these matrices we can write the Larkin equation similar to the Fermi-type quasiparticles,

$$\hat{D}(k,\omega_n) = -[\hat{T}^{-1}(k,\omega_n) + \hat{V}_k]^{-1}. \tag{5.79}$$

The split (multiplicated) form of the interaction matrix again allows one to simplify the solution of the Larkin equation. However for Bose-particles some differences resulted from several types of interaction lines. To clarify the origin of this peculiarity we consider for example the exchange interaction of the Heisenberg type

$$-\frac{1}{2}\sum_{fg} I_{fg}(\boldsymbol{S}_f \boldsymbol{S}_g) = -\frac{1}{2}\sum_{fg} I_{fg}\left(\frac{1}{2}S_f^+ S_g^- + \frac{1}{2}S_f^- S_g^+ + S_f^z S_g^z\right)$$

This type of interaction takes place not only in magnetic insulators but also in systems with mobile carriers such as the t–J model. In general interaction of Bose-type contains product of non-Hermitian[1] operators as well as Hermitian (here S^z) operators c_f and it may be written as

$$H_{\text{int}}^{(B)} = \sum_{fg}\sum_{ij} t_{fg}^{ij} B_{fi}^+ B_{gj} + \frac{1}{2}\sum_{fg}\sum_{rs} u_{fg}^{rs} C_{fr} C_{gs}. \tag{5.80}$$

In the Hubbard representation of the B and C operators, there are in general both non-diagonal and diagonal X-operators,

$$B_{fi} = \sum_\lambda \gamma_i(\lambda) X_f^\lambda, \qquad C_{fr} = \sum_\lambda p_r(\lambda) X_f^\lambda. \tag{5.81}$$

Here, index λ labels both the root vectors α and the intracell eigenstates n; indices r, s numerate the intracell eigenstates n. It is evident that Hermitian

---

[1] In this example it is S^+, S^-.

conjugation results in

$$t_{fg}^{ij} = \left(t_{gf}^{ji}\right)^*, \qquad \left(u_{fg}^{rs}\right)^* = u_{fg}^{sr} = u_{gf}^{sr}, \qquad p_r(\lambda) = (p_r(-\lambda))^*.$$

We introduce the following vector

$$C(\lambda) = \{\gamma_1(\lambda), \gamma_2(\lambda), \ldots, \gamma_1^*(-\lambda), \gamma_2^*(-\lambda), \ldots, p_1(\lambda), p_2(\lambda), \ldots\}, \quad (5.82)$$

and the block matrix

$$\hat{V}_{fg} = \begin{pmatrix} \hat{t}_{fg} & 0 & 0 \\ 0 & \hat{t}_{fg}^* & 0 \\ 0 & 0 & \hat{u}_{fg} \end{pmatrix},$$

with matrix elements t_{fg}^{ij}, $\left(t_{fg}^{ij}\right)^*$ and u_{fg}^{rs}. In this notation the interaction of Bose-type can be written as

$$H_{\text{int}}^{(B)} = \frac{1}{2} \sum_{fg} \sum_{\lambda\lambda'} \{C(-\lambda), \hat{V}_{fg} C(\lambda')\} X_f^\lambda X_g^{\lambda'},$$

i.e. similar to the Fermi-type interaction with the only difference in the index λ. This similarity yields solution similar to that of the Larkin equation for the Bose-type Green's function

$$D_{\lambda\lambda'} = -T_{\lambda\lambda'} - \sum_{\lambda_1} T_{\lambda\lambda_1}\{C(-\lambda_1), [V_k^{-1} + L]^{-1} B_\lambda\}, \quad (5.83)$$

$$L_{ij}(\mathbf{k}, \omega_n) = \sum_{\lambda_1\lambda_2} C_i(\lambda_1) T_{\lambda_1\lambda_2}(\mathbf{k}, \omega_n) C_j(-\lambda_2). \quad (5.84)$$

One more restriction for the Bose quasiparticles is the absence of Bose condensate. In principal there is no restriction to generalize X-diagram technique to Bose condensate systems similar to the usual Bose type operators but it should be done separately and is outside of the scope of this book.

5.10 The Self-Energy and the Strength Operator

In the conventional diagram technique it is convenient to introduce the self-energy part of the Green's function and to right down the Dyson equation. It occurs that for X-operators we can also extract the self-energy part $\sum_{\lambda\lambda'}$ from all irreducible Larin diagrams $T_{\lambda\lambda'}$. For spin operators diagram technique this result have been obtained by Baryakchtar, Yablonsky and Kryvoruchko (1983) and then used on the X-operators by Izyumov and Letfullov (1991).

For simplicity we shall restrict ourselves in this chapter by considering only one type of interaction line — the transverse interaction of the Fermi type. One of the physical examples for such a case is given by the Hubbard model in the $U = \infty$ limit, with the Hamiltonian

$$H - \mu \hat{N} = \sum_{f\sigma}(\varepsilon_\sigma - \mu)X_f^{\sigma\sigma} + \sum_{fg}\sum_\sigma t_{fg}X_f^{\sigma 0}X_g^{0\sigma}. \qquad (5.85)$$

We can write the interaction part of the Hamiltonian in the form

$$H_{\text{int}} = \sum_{fg}\sum_m t_{fg}X_f^{m0}X_g^{0m} = \frac{1}{2}\sum_{fg}\sum_{\alpha\beta}V_{fg}^{\alpha\beta}X_f^\alpha X_g^\beta \qquad (5.86)$$

with the interaction matrix given by

$$V_{fg}^{\alpha\beta} = \begin{cases} t_{fg}\delta_{\alpha,-\beta}\text{sign}(\beta) & \alpha \text{ is of Fermi-type} \\ 0 & \text{otherwise} \end{cases}. \qquad (5.87)$$

Here we ascribe $\text{sign}(\beta) = +1$ if the Fermi root vector β corresponds to electron annihilation, and $\text{sign}(\beta) = -1$ otherwise. The condition (5.87) is not really used below so the consideration is valid for arbitrary transverse interaction.

Let us consider several diagrams for $T_{\alpha\beta}$ (5.88):

$$(5.88)$$

where the Fermi (Bose) type Green's functions are shown by solid (dotted) lines. The anticommutation of two Fermi type X-operators results in the

Bose type X-operator, thus in the nondiagonal three-leg vertex there is a change of the Fermi type quasiparticle to the Bose type.

All the diagrams in the graph (5.88) and, in general in $T_{\alpha\beta}$, can be subdivided into three classes with different way of connecting the entrance point and the exit point:

1. The first class is given by diagrams where the entrance point and the exit point are connected with only one propagator line (graphs 1, 2, 3 in (5.88));
2. The second class is formed by diagrams where the entrance and exit points are connected with more then one propagator in such a way that if one wish to separate the entrance and exit point by cutting one propagator line one can only do it if this propagator line starts from the entrance point (graphs 4 and 5 in (5.88));
3. The third class is given by diagrams that can be cut by propagator line that does not starts from the entrance point (6, 7, 8 in (5.88)).

The sum of all the first type diagrams can be written as follows

$$ \text{(diagrammatic equation)} \tag{5.89} $$

where the full circle is given by

$$ \text{(diagrammatic equation)} \tag{5.90} $$

Thus high-order renormalization transforms an empty circle to the full circle in a way similar to the case of the spin operator diagram technique (Izyumov, Katsnelson and Skryabin, 1994).

The sum of all the second type diagrams will be denote by a full triangle

$$\text{[diagram]} \quad (5.91)$$

We introduce a new element, a full semicircle ▶, which is equal to

$$\text{[diagram]}$$

It is clear that the full semicircle is the exact two-leg vertex P. In the spin diagram technique it was called "a strength operator" (Baryakchtar, Yablonsky and Kryvoruchko 1983) because P is in the numerator of the Green's function and the value of P determines an oscillator strength of excitations.

Now we can analyze diagrams of the third class. The full set of diagrams for $T_{\alpha\beta}$ after partial summation can be written as follows

$$\text{[diagram]} \quad (5.92)$$

Here the second and the third graphs consist of some fragments connected by two propagator lines. These fragments look like

$$\text{[diagram]} \quad (5.93)$$

Similarly, the forth graph in (5.92) is given by a fragment

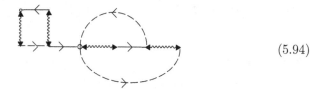
(5.94)

that consists of two more simple fragments (5.93) connected by one propagator line.

As usually in diagram technique, a fragment connected with other parts of diagram with two propagators is the self-energy part of diagram. The graph (5.93) shows two irreducible self-energy parts while the graph (5.94) shows the reducible self-energy part, meaning that reducible part can be cut over the propagator line. A sum of all irreducible self-energy diagrams is called the self-energy part Σ of Green's function, it will be shown by

 .

Some diagrams for the self-energy are given by

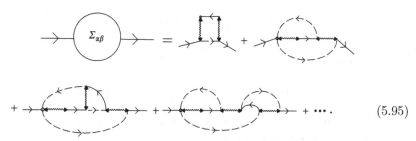
(5.95)

The self-energy and the strength operator determine the sum of diagrams for $T_{\alpha\beta}$

$$\left(T\right) = \rightarrow\!\!\blacktriangleright + \rightarrow\!\!\left(\Sigma\right)\!\!\rightarrow\!\!\blacktriangleright$$

$$+ \rightarrow\!\!\left(\Sigma\right)\!\!\rightarrow\!\!\left(\Sigma\right)\!\!\rightarrow\!\!\blacktriangleright + \cdots. \qquad (5.96)$$

Analytical expression reads as

$$T_{\alpha\beta} = -G_\alpha P_{\alpha\beta} + G_\alpha \Sigma_{\alpha\beta_1} G_{\beta_1} P_{\beta_1\beta} - G_\alpha \Sigma_{\alpha\beta_1} G_{\beta_1} \Sigma_{\beta_1\beta_2} G_{\beta_2} P_{\beta_1\beta} + \cdots. \tag{5.97}$$

We introduce matrices $\hat{T}$ and $\hat{P}$ by $\left(\hat{T}\right)_{\alpha\beta} = T_{\alpha\beta}$ and $\left(\hat{P}\right)_{\alpha\beta} = P_{\alpha\beta}$, then

$$\hat{T} = -\left(\hat{G}_0^{-1} + \hat{\Sigma}\right)^{-1}\hat{P}, \tag{5.98}$$

where the matrix $\hat{G}_0$ is given by

$$(G_0)_{\alpha\beta} = \delta_{\alpha\beta} G_\alpha(\omega_n). \tag{5.99}$$

Finally the Larkin equation (5.68) yields

$$\hat{D}(\bm{q},\omega_n) = \left[\hat{G}_0^{-1}(\omega_n) + \hat{\Sigma}(\bm{q},\omega_n) - \hat{P}(\bm{q},\omega_n)\hat{V}_{\bm{q}}\right]^{-1}\hat{P}(\bm{q},\omega_n). \tag{5.100}$$

To write this equation in the form of the Dyson equation we introduce the simplest delocalized Green's function $\hat{G}^{(0)}(\bm{q},\omega_n)$ given by

$$\Longrightarrow = \longrightarrow + \longrightarrow\blacktriangleleft\!\!\sim\!\!\sim\!\!\sim\!\!\blacktriangleright\!\!\Longrightarrow \tag{5.101}$$

Analytical expression of this equation can be written as

$$\hat{G}^{(0)}(\bm{q},\omega_n) = \hat{G}_0(\omega_n) + \hat{G}_0(\omega_n)P(\bm{q},\omega_n)V_{\bm{q}}\hat{G}^{(0)}(\bm{q},\omega_n).$$

In matrix form, this Green's function is given by

$$\left[\hat{G}^{(0)}(\bm{q},\omega_n)\right]^{-1} = \left[\hat{G}_0(\omega_n)\right]^{-1} - P(\bm{q},\omega_n)V_{\bm{q}}. \tag{5.102}$$

Using the function $G^{(0)}$ we can write the exact equation (5.100) for $\hat{D}$ in the form of the generalized Dyson equation

$$\hat{D}(\bm{q},\omega_n) = \left\{\left[\hat{G}^{(0)}(\bm{q},\omega_n)\right]^{-1} + \hat{\Sigma}(\bm{q},\omega_n)\right\}^{-1}\hat{P}(\bm{q},\omega_n). \tag{5.103}$$

Contrary to the conventional Dyson equation where the numerator of the full Green's function is equal to 1 and the renormalization of the quasiparticle spectral weight stems only from the self-energy contribution like in the Fermi-liquid theory (Abrikosov, Gor'kov and Dzyaloshinsky 1965), in the strongly correlated systems a new notion of the strength operator appears in the numerator of the full Green's function that determines the spectral weight of the quasiparticle. We want to emphasize that the origin of the strength operator is the algebra of X-operators, particularly the anticommutation rule

$$X^{pq}X^{qp} + X^{qp}X^{pq} = X^{pp} + X^{qq} \neq 1.$$

Thus the fractionalization of electron in strongly correlated systems in our approach means the fractionalization of the quasiparticle spectral weight with electron charge e and spin $S = 1/2$ in each excitation channel determined by the Fermi-type root vector $\boldsymbol{\alpha}$.

To use the generalized Dyson equation (5.103) in the perturbation expansion it is necessary to calculate both functions $\hat{\Sigma}(\boldsymbol{q},\omega_n)$ and $\hat{P}(\boldsymbol{q},\omega_n)$ in the same order of perturbation. We give such example in the next section.

5.11 Low Temperature Properties of Anisotropic Ferromagnet

We consider here for illustration of the diagram technique the Heisenberg ferromagnet with a single axis anisotropy in the external magnetic field H. The simplest geometry takes place when H and the magnetization σ are both directed along the anisotropy axis. It may be for the easy axis anisotropy at arbitrary value of magnetic field H and for the easy plane anisotropy for the field $H > H_c$, where the value of critical field H_c is given below in the Chapter 6.

The Hamiltonian of this ferromagnet is given by

$$H = -\frac{1}{2}\sum_{fm} I_{fm}(\boldsymbol{S}_f \boldsymbol{S}_m) - \sum_f \varphi(S_f^z) - g\mu_B H \sum_f S_f^z. \quad (5.104)$$

Here $\boldsymbol{S}_f$ is the spin operator at site f with components S_f^x, S_f^y, S_f^z. An operator function

$$\varphi(S_f^z) = D_2(S_f^z)^2 + D_4(S_f^z)^4 \quad (5.105)$$

describes the single-ion anisotropy. To construct X-operators we introduce a set of local eigenstates of the S_f^z operator

$$S_f^z|f,M\rangle = M|f,M\rangle, \quad M = S, S-1, S-2, \ldots, -S+1, -S \quad (5.106)$$

and define the X-operators as

$$X_f^{M_1 M_2} \equiv |f,M_1\rangle\langle M_2,f|.$$

In the X-operator representation the spin operators are equal to

$$S_f^+ = \sum_{M=-S}^{S} \gamma_S(M) X_f^{M+1,M}, \qquad S_f^- = \sum_{M=-S}^{S} \gamma_S(M) X_f^{M,M+1},$$
$$S_f^z = \sum_M M X_f^{MM}, \qquad \gamma_S(M) = \sqrt{(S-M)(S+M+1)}.$$
(5.107)

We can write this representation as a sum over all Bose type root vectors

$$S_f^+ = \sum_\alpha \gamma(\alpha) X_f^\alpha, \qquad S_f^- = \sum_\alpha \gamma(\alpha) X_f^{-\alpha},$$
(5.108)

by introducing a function $\gamma(\alpha)$,

$$\gamma(\alpha) = \begin{cases} \gamma_S(M), & \alpha = \alpha(M+1, M), \quad -S \le M \le S-1, \\ 0, & \alpha \ne \alpha(M+1, M) \end{cases}$$

Thus the Hamiltonian (5.104) in the X-representation looks like

$$H = \frac{1}{2} N I_0 \sigma^2 + H_0 + H_{\text{int}}.$$
(5.109)

Here the magnetization is equal to

$$\sigma = \frac{1}{N} \sum_f \langle S_f^z \rangle.$$
(5.110)

The intracell part of the Hamiltonian is given by

$$H_0 = \sum_f \sum_{M=-S}^{S} E_M X_f^{MM}, \qquad E_M = -\bar{H} M - \varphi(M),$$
(5.111)

and corresponds to a sum of N magnetic ions with non-equidistant energy levels E_M in the mean field $\bar{H}$,

$$\bar{H} = g\mu_B H + I_0 \sigma, \qquad I_0 = \sum_h I(h).$$
(5.112)

The intercell part of the Hamiltonian H_{int} contains

$$H_{\text{int}} = -\frac{1}{2} \sum_{fm} \sum_{\alpha\beta} V_{fm}^{\alpha\beta} X_f^\alpha X_m^\beta - \frac{1}{2} \sum_{fm} \sum_{MM'} I_{fm}^{MM'} \Delta X_f^{MM} \Delta X_m^{M'M'},$$
(5.113)

both transverse and longitudinal interactions

$$V_{fm}^{\alpha\beta} = I_{fm}\frac{1}{2}[\gamma(\alpha)\gamma(-\beta) + \gamma(-\alpha)\gamma(\beta)],$$

$$I_{fm}^{MM'} = I_{fm}MM', \qquad \Delta X_f^{MM} = X_f^{MM} - \sigma.$$

In the low temperature limit $T \ll T_c$, where T_c is the Curie temperature, an anisotropy energy ε_a and Zeeman splitting are both small in comparison with the exchange energy

$$g\mu_B H \approx \varepsilon_a \ll I_0 S.$$

It means that population of all excited single-ion energy levels is negligibly small except for the ground state with $M = S$. Thus only one local excitation $|S\rangle \to |S-1\rangle$ will be dispersed due to the interactions in the one spin wave band.

In the spin diagram technique the self-consistent field (SCF) approximation is known where the Green's function can be calculated in the loop expansion (Vaks, Larkin and Pikin 1967a) with each loop being proportional to the inverse volume of the interaction $1/r_0^3$. The zeroth order SCF approximation which has no loop is equal to the Hartree–Fock approximation

$$\Rrightarrow = \to + \to\!\!\circ\!\!\leftsquigarrow\!\!\blacktriangleright\!\!\Rightarrow. \qquad (5.114)$$

Here a double line with arrows denotes the Hartree–Fock Green's function $G_{\alpha\beta}^{(0)}$, thin solid line with arrow means a bare propagator $G_\alpha(\omega_n) = [i\omega_n + \alpha E]^{-1}$, and wavy line with two arrows denotes the transverse interaction

$$V_q^{\alpha\beta} = \frac{1}{2}I_q(\gamma(\alpha)\gamma(-\beta) + \gamma(-\alpha)\gamma(\beta)). \qquad (5.115)$$

An empty circle in Eq. (5.114) means the strength operator. In the zeroth order SCF approximation the latter is equal to

$$b(\alpha) = N_M - N_{M'} \quad \text{when} \quad \alpha = \alpha(M, M'). \qquad (5.116)$$

For the root vectors $\gamma(\alpha)$ with $\alpha = (M, M \pm 1)$, we have

$$b(\alpha) = N_M - N_{M\pm 1}.$$

Thus neglecting exponentially small occupation $\sim e^{-\Delta E/T}$ of the excited single-ion levels at low T we get

$$b(\alpha_0) = 1, \quad b(-\alpha_0) = -1, \quad \alpha_0 \equiv \alpha(S, S-1), \qquad (5.117)$$

and $b(\alpha) = 0$ for all other vectors α. The analytical expression of Eq. (5.114) is written as follows

$$G^{(0)}_{\alpha\beta}(k,\omega_n) = \delta_{\alpha\beta} G_\alpha(\omega_n) - \sum_{\beta_1} G_\alpha(\omega_n) b(\alpha) V_k^{-\alpha\beta_1} G^{(0)}_{\beta_1\beta}(k,\omega_n). \quad (5.118)$$

At $T \ll T_c$ due to condition (5.117), there is no dispersion for excitations with $\alpha \neq \alpha_0$ and $\alpha \neq -\alpha_0$,

$$G^{(0)}_{\alpha\beta}(k,\omega_n) = \delta_{\alpha\beta} G_\alpha(\omega_n). \quad (5.119)$$

For the spin wave excitation with $\alpha = \alpha_0$, Eq. (5.118) yields

$$(i\omega_n - \omega_k) G^{(0)}_{\alpha_0\beta}(k,\omega_n) = \delta_{\alpha_0\beta} - \frac{\sqrt{2S}}{2}\gamma(\beta)(1-\delta_{\alpha_0\beta}) I_k G_\beta(\omega_n), \quad (5.120)$$

with magnon spectrum

$$\omega_k = g\mu_B H + \varepsilon_a + S(I_0 - I_k), \quad \varepsilon_a = \varphi(S) - \varphi(S-1). \quad (5.121)$$

The zeroth order Green's functions are equal to

$$G^{(0)}_{\alpha_0\alpha_0}(k,\omega_n) = \frac{1}{i\omega_n - \omega_k},$$

$$G^{(0)}_{\alpha_0\beta}(k,\omega_n) = -\frac{\sqrt{2S}\gamma(\beta)}{2} I_q \left(\frac{1}{i\omega_n - \omega_k}\right)\left(\frac{1}{i\omega_n + \beta E}\right), \quad \beta \neq \alpha_0. \quad (5.122)$$

Now we will discuss the correlation effects. It is convenient here to specify the root vector α in such a manner that it corresponds to the excitations with spin projection increased by one. In other words, it means the annihilation of magnon. Then we can write the transverse interaction in the form

$$H_{\text{int}} = -\frac{1}{2}\sum_{fm}\sum_{\alpha\beta} V^{\alpha\beta}_{fm} X_m^{-\beta} X_f^\alpha, \quad V^{\alpha\beta}_{fm} = \frac{1}{2} I_{fm}\gamma(\alpha)\gamma(\beta). \quad (5.123)$$

In this notation the interaction line is not symmetric in arrows, it look like

$$\underset{\beta \qquad \alpha}{\blacktriangleright\!\sim\!\sim\!\sim\!\blacktriangleright}$$

and has the entrance and exit arrows.

In the first order term over $1/r_0^3$ in the SCF approximation we have to consider one-loop diagrams. One-loop diagrams for the strength operator $P_{\alpha\beta}$ look like

$$\tag{5.124}$$

We simplify notations as follow the root vector α_0 has index 1, $\alpha_1 \equiv \alpha(S-1, S-2)$ is marked by index 2, etc. Then the left graph (5.124) gives only two non-zero components

$$P_{11} = -2\bar{n}, \qquad P_{22} = \bar{n}, \tag{5.125}$$

where $\bar{n}$ is a magnon concentration

$$\bar{n} = \frac{1}{N}\sum_q n_q = \frac{1}{N}\sum_q \frac{1}{\exp(\omega_q/T) - 1}.$$

To obtain the expression (5.125) we have used two factors. Firstly, the empty circle gives $b(\beta_1)$ and the only non-zero contribution here is given by $\beta_1 = \alpha_0 = \alpha(S, S-1)$. Secondly, the four-leg pricked circle vertex is determined by a scalar product of two root vectors

$$(\alpha\beta_1) = (\alpha\alpha_0) = \sum_n (\alpha)_n (\alpha_0)_n = \sum_n (\alpha)_n (\delta_{nS} - \delta_{n,S-1}).$$

The nonzero components of this product appear when $\alpha = \alpha_0$ and $\alpha = \alpha_1$. The former case yields 2 and the latter gives (-1).

Similar calculation of the right graph (5.124) results in

$$P_{12} = -\frac{1}{N}\sum_q \frac{\sqrt{S(2S-1)}I_q}{i\omega_n + \omega_q + E_S - E_{S-2}} n_q,$$

$$P_{22} = \frac{1}{N}\sum_q \frac{SI_q}{i\omega_n + \omega_q + E_S - E_{S-2}} n_q.$$

Finally, the first order SCF contribution to the strength operator is equal to

$$P_{11} = -\frac{2}{N}\sum_q n_q,$$

$$P_{12} = -\frac{1}{N}\sum_q \frac{\sqrt{S(2S-1)}I_q}{i\omega_n + \omega_q + E_S - E_{S-2}}n_q, \qquad (5.126)$$

$$P_{22} = \frac{1}{N}\sum_q \left\{1 + \frac{SI_q}{i\omega_n + \omega_q + E_S - E_{S-2}}\right\}n_q.$$

Now we can calculated the Hartree–Fock Green's function using Eq. (5.101). For functions $G_{11}^{(0)}(\boldsymbol{k},\omega_n)$ and $G_{21}^{(0)}(\boldsymbol{k},\omega_n)$ one has the next system of equations

$$G_{11}^{(0)} = G_{\alpha_0} - G_{\alpha_0}\left(P_{11}V^{11} + P_{12}V^{21}\right)G_{11}^{(0)} - G_{\alpha_0}\left(P_{11}V^{12} + P_{12}V^{22}\right)G_{21}^{(0)},$$
$$G_{21}^{(0)} = -G_{\alpha_1}P_{22}V^{21}G_{11}^{(0)} - G_{\alpha_1}P_{22}V^{22}G_{21}^{(0)}, \qquad (5.127)$$

one solution of Eq. (5.127) is given by

$$G_{11}^{(0)}(\boldsymbol{k},\omega_n) = \frac{i\omega_n + \alpha_1 E + P_{22}V_k^{22}}{\Delta_0}, \qquad G_{21}^{(0)}(\boldsymbol{k},\omega_n) = -\frac{P_{22}V_k^{21}}{\Delta_0},$$

$$\Delta_0 = \left(i\omega_n + \alpha_0 E + P_{11}V_k^{11} + P_{12}V_k^{21}\right)\left(i\omega_n + \alpha_1 E + P_{22}V_k^{22}\right)$$
$$- P_{22}V_k^{21}\left(P_{11}V_k^{12} + P_{12}V_k^{22}\right). \qquad (5.128)$$

In the same manner we calculate another zeroth-order Green's functions

$$G_{22}^{(0)}(\boldsymbol{k},\omega_n) = \frac{i\omega_n + \alpha_0 E + P_{11}V_k^{11} + P_{12}V_k^{21}}{\Delta_0},$$

$$G_{12}^{(0)}(\boldsymbol{k},\omega_n) = -\frac{P_{11}V_k^{12} + P_{12}V_k^{22}}{\Delta_0}.$$

The Dyson equation now looks like

$$\longrightarrow\!\!\!\longrightarrow \;=\; \Longrightarrow\!\!\!\Longrightarrow \;+\; \Longrightarrow\!\!\!\Longrightarrow\!\!\boxed{\Sigma}\!\!\longrightarrow\!\!\!\longrightarrow. \qquad (5.129)$$

with the self-energy given in the one-loop approximation by a sum of three graphs arising from magnon–magnon scattering

$$\qquad (5.130)$$

Here bold wavy line denotes an effective interaction that is found from

and is equal to

$$\tilde{V}_q^{\alpha\beta}(\omega_n) = \frac{i\omega_n + \alpha_0 E}{i\omega_n - \omega_q} V_q^{\alpha\beta}, \quad V_q^{\alpha\beta} = \frac{1}{2}\gamma(\alpha)I_q\gamma(\beta).$$

For the ground state excitation the effective interaction is given by

$$\tilde{V}_q^{\alpha_0\alpha_0}(\omega_n) = \frac{i\omega_n + \alpha_0 E}{i\omega_n - \omega_q} SI_q.$$

Analytical expression for the Dyson equation (5.129) yields for the Green's function G_{11} the following equation

$$G_{11}(\bm{k},\omega_n) = G_{11}^{(0)}(\bm{k},\omega_n) - G_{11}^{(0)}(\bm{k},\omega_n)\Sigma_{11}(\bm{k},\omega_n)G_{11}(\bm{k},\omega_n)$$
$$- G_{12}^{(0)}(\bm{k},\omega_n)\Sigma_{21}(\bm{k},\omega_n)G_{11}(\bm{k},\omega_n) \tag{5.131}$$

with a solution given by

$$G_{11} = \left\{ i\omega_n + \alpha_0 E + P_{11}V^{11} + P_{12}V^{21} + \Sigma_{11} - \frac{P_{22}V^{21}V^{12} + V^{12}\Sigma_{21}}{i\omega_n + \alpha_1 E} \right\}^{-1}. \tag{5.132}$$

A calculation of the self-energy contribution in Eq. (5.130) results in

$$\Sigma_{11} = \frac{1}{N}\sum_q (I_{\bm{k}-\bm{q}} - I_q)n_q - \frac{(2S-1)}{N}\sum_q I_q \frac{(i\omega_n + \alpha_1 E)}{(i\omega_n + \omega_q + E_S - E_{S-2})}n_q,$$
$$\Sigma_{21} = \sqrt{S(2S-1)}\frac{1}{N}\sum_q \frac{(i\omega_n + \alpha_1 E)}{(i\omega_n + \omega_q + E_S - E_{S-2})}I_q n_q. \tag{5.133}$$

Finally, the denominator of the Green's function G_{11} can be written as

$$\Delta(\bm{k},\omega_n) = i\omega_n - \omega_{\bm{k}} + \frac{1}{N}\sum_q (I_0 - I_{\bm{k}} + I_{\bm{k}-\bm{q}} - I_q)n_q$$
$$- \frac{(2S-1)}{N}\sum_q \frac{(I_{\bm{k}} + I_q)(i\omega_n + SI_{\bm{k}} + \alpha_1 E)}{i\omega_n + \omega_q + E_S - E_{S-2}}n_q.$$

Close to bare magnon energy $i\omega_n \approx \omega_k$, the latter can be simplified as follows

$$\Delta(k,\omega_n) = i\omega_n - \omega_k + \frac{1}{N}\sum_q (I_0 - I_k + I_{k-q} - I_q)n_q$$

$$+ \frac{(2S-1)R[\varphi]}{N}\sum_q \frac{I_k + I_q}{i\omega_n + \omega_q + E_S - E_{S-2}}n_q,$$

where the anisotropic factor $R[\varphi]$ is equal to

$$R[\varphi] = 2\varphi(S-1) - \varphi(S) - \varphi(S-2).$$

Finally, the renormalized magnon energy is given by

$$\Omega_k = \omega_k - \frac{1}{N}\sum_q (I_0 - I_k + I_{k-q} - I_q)n_q$$

$$- \frac{(2S-1)R[\varphi]}{N}\sum_q \frac{I_k + I_q}{\omega_k + \omega_q + E_S - E_{S-2}}n_q.$$

Here the first sum over q is the usual Dyson term for the isotropic ferromagnet while the last sum results from anisotropy. The latter can be written as follows

$$\delta^{(a)}\omega_k = -\frac{(2S-1)R[\varphi](I_0 + I_k)}{S(I_0 + I_k) + R[\varphi]}\frac{1}{N}\sum_q n_q$$

$$+ \frac{(2S-1)R^2[\varphi]}{S(I_0 + I_k) + R[\varphi]}\frac{1}{N}\sum_q (I_0 - I_q)n_q.$$

At $T \ll T_c$ the first term here gives the main contribution. We denote the effective radius of interaction as

$$r_0^2 = \frac{\sum_h h^2 I(h)}{\sum_h I(h)}$$

and obtain

$$\delta^{(a)}\omega_k = -\frac{(2S-1)R[\varphi](I_0 + I_k)}{S(I_0 + I_k) + R[\varphi]}\left(\frac{a}{r_0}\right)^3\left(\frac{T}{4\pi SI}\right)^{3/2} Z_{3/2}(\Delta/T).$$

Here a is a lattice parameter (we assume here a SC lattice),

$$I = \frac{1}{6}\sum_h I(h).$$

A gap in the spin-wave spectrum is equal to

$$\Delta = g\mu_B H + \varepsilon_a.$$

For the easy plane ferromagnet $\varepsilon_a < 0$ and the value of Δ is comparable to or even smaller than temperature. Then the anisotropy renormalization of magnon spectrum $\sim (T/T_c)^{3/2}$ can exceed the Dyson renormalization $\sim (T/T_c)^{5/2}$.

Similarly we can calculate a temperature dependence of the magnetization due to magnon–magnon scattering. In the isotropic ferromagnet it results in $(T/T_c)^4$ contribution (Dyson 1956). In the single-ion anisotropic ferromagnet the temperature dependence of magnon induced by the magnon–magnon scattering has the form (Val'kov and Ovchinnikov 1983).

$$\delta S^z = \delta^{(D)} S^z + \delta^{(a)} S^z. \tag{5.134}$$

Here $\delta^{(D)} S^z$ is the Dyson contribution

$$\delta^{(D)} S^z = -\left(\frac{1}{r_0^3}\right)^2 \frac{\pi\mu}{6S} \left(\frac{3T}{2\pi S I_0}\right)^4 Z_{5/2}\left(\frac{\Delta}{T}\right) Z_{3/2}\left(\frac{\Delta}{T}\right), \tag{5.135}$$

and the anisotropy contribution looks like

$$\delta^{(a)} S^z = \left(\frac{1}{r_0^3}\right)^2 \frac{3(2S-1)}{\pi S} \frac{R[\varphi]}{2S I_0 + R[\varphi]} \left(\frac{3T}{2\pi S I_0}\right)^2 \left\{ Z_{3/2}\left(\frac{\Delta}{T}\right) Z_{1/2}\left(\frac{\Delta}{T}\right) \right.$$
$$\left. -\frac{\pi}{2} \frac{R[\varphi]}{2S I_0 + R[\varphi]} \left(\frac{3T}{2\pi S I_0}\right) \left[Z_{3/2}^2\left(\frac{\Delta}{T}\right) + Z_{1/2}\left(\frac{\Delta}{T}\right) Z_{5/2}\left(\frac{\Delta}{T}\right)\right] \right\}.$$
$$\tag{5.136}$$

This contribution tends to zero while its anisotropy contribution may exceed the Dyson one. For example, if $\varphi(S_f^z) = D(S_f^z)^2$ then at $T \leq \Delta < I_0 S$ we obtain

$$\delta^{(D)} S^z = -\left(\frac{1}{r_0^3}\right)^2 \frac{\pi\mu}{6S} \left(\frac{3T}{2\pi S I_0}\right)^4 \exp(-2\Delta/T),$$

$$\delta^{(a)} S^z = -\left(\frac{1}{r_0^3}\right)^2 \frac{3(2S-1)}{\pi S} \frac{D}{S I_0 - D} \left(\frac{3T}{2\pi S I_0}\right)^2 \exp(-2\Delta/T). \tag{5.137}$$

It is clear that at low temperatures

$$\delta^{(D)} S^z \ll \delta^{(a)} S^z.$$

At higher temperatures $\Delta \ll T \ll I_0 S$, we get

$$\delta^{(D)} S^z \sim -\frac{1}{Sr_0^6}\left(\frac{T}{SI_0}\right)^4, \qquad \delta^{(a)} S^z \sim -\frac{2S-1}{Sr_0^6}\left(\frac{D}{SI_0}\right)\left(\frac{T}{SI_0}\right)^2 \left(\frac{T}{\Delta}\right)^{1/2}$$

and again the anisotropy contribution is larger than Dyson one.

The strongest effect of the anisotropy takes place when $D < 0$ (easy plane anisotropy), and the external magnetic field is large enough to stabilize the colinear magnetic order. It requires $g\mu_B H > |D|(2S-1)$. In the case of rather strong anisotropy $|D|$ one can find such value of the magnetic field that the $Z_{1/2}(\Delta/T)$ function will not be small. In this situation magnetization which decreases due to anisotropy $\delta^{(a)} S^z$ may be quite large, more than 10% from the magnon Bloch contribution.

5.12 Effect of the Three-Site Correlated Hopping on the Magnetic Mechanism of $d_{x^2-y^2}$-Superconductivity in the t–J^* Model

It is well known that the t–J model (Anderson 1987) qualitatively correctly describes magnetic mechanism of pairing in HTSC cuprates (see e.g. a review by Izyumov (1997)). This model is obtained from the Hubbard model in the SEC regime $t \ll U$ as the effective model excluding double occupation of local states. The effective Hamiltonian (see Chapter 1 above) has more a general form $H_{\text{eff}} = H_{t-J} + H_{(3)}$, where $H_{(3)}$ corresponds to correlated three-site hopping (Bulaevskii, Nagaev and Khomskii 1968; Chao, Spalek and Oles 1977). Yushankhay, Oudovenko and Hayn (1997) has shown that the effect of $H_{(3)}$ on the quasiparticle spectra is small because the amplitude of correlated three-site hopping $t^2/U \ll t$. For magnetic mechanism of pairing the effect of $H_{(3)}$ may be more important because the coupling constant J is of the same order of magnitude, i.e. $J \sim t^2/U$. This effect was first studied by Hirsch (1989). Yushankhay, Vujicic and Zakula (1990) has shown that $H_{(3)}$ contribution renormalizes the effective coupling constant in the gap equation. Recently Val'kov et al (2002) has found that the effect of $H_{(3)}$ may strongly suppress T_c of the magnetic mechanism of pairing. We will describe this effect below.

The effective Hamiltonian is derived from the Hubbard model and acts in the Hilbert subspace with constraint of no doubly occupied states. We will derive it in the X-representation. In the X-representation the Hubbard model Hamiltonian looks like $H = H_0 + V$, where H_0 is the sum of all local energies and V includes all hopping terms (see Eq. (1.51)). With an accuracy quadratic in t_{fm}/U, the effective Hamiltonian is determined by the equation

$$H_{eff} = PH_0P + PVP + PV(H_0 - E_0)^{-1}(PVP - VP), \qquad (5.138)$$

where P is the operator of projection onto the Hilbert subspace of no doubly occupied state. Using the multiplication rules for Hubbard operators $X_f^{nm} X_f^{lq} = \delta_{ml} X_f^{nq}$, we obtain

$$P = \prod_f \left(X_f^{00} + X_f^{\uparrow\uparrow} + X_f^{\downarrow\downarrow} \right), \qquad PVP - VP = -\sum_{fm\sigma} t_{fm} \eta(\sigma) X_m^{2\bar{\sigma}} X_f^{0\sigma} P. \qquad (5.139)$$

With regard to these relationships, we find the third term in H_{eff} to be

$$PV(H_0 - E_0)^{-1}(PVP - VP)$$
$$= P \sum_{fmg\sigma} \left(\frac{t_{fm} t_{mg}}{U} \right) \left(X_f^{\sigma 0} X_m^{\bar{\sigma}\sigma} X_g^{0\bar{\sigma}} - X_f^{\sigma 0} X_m^{\bar{\sigma}\bar{\sigma}} X_g^{0\sigma} \right) P. \qquad (5.140)$$

Addition of the two first terms from Eq. (5.138) leads to the following structure of H_{eff}:

$$H_{eff} = \sum_{f\sigma} (\varepsilon - \mu) X_f^{\sigma\sigma} + \sum_{mf\sigma} t_{fm} X_f^{\sigma 0} X_m^{0\sigma}$$
$$+ \sum_{fmg\sigma} \left(\frac{t_{fm} t_{mg}}{U} \right) \left(X_f^{\sigma 0} X_m^{\bar{\sigma}\sigma} X_g^{0\bar{\sigma}} - X_f^{\sigma 0} X_m^{\bar{\sigma}\bar{\sigma}} X_g^{0\sigma} \right).$$

In this equation, the projection operator is omitted, because the Hilbert subspace without doubly occupied state is invariant with respect to the action of H_{eff}. In other words in this subspace $P = 1$. Among others, the last summand in H_{eff} contains terms with $f = g$, which, taken together with the two first terms of H_{eff}, give the Hamiltonian of the $t-J$ model

$$H_{t-J} = \sum_{f\sigma} (\varepsilon - \mu) X_f^{\sigma\sigma} + \sum_{f\sigma} t_{fm} X_f^{\sigma 0} X_m^{0\sigma}$$
$$+ \frac{1}{2} \sum_{fm\sigma} J_{fm} \left(X_f^{\sigma\bar{\sigma}} X_m^{\bar{\sigma}\sigma} - X_f^{\sigma\sigma} X_m^{\bar{\sigma}\bar{\sigma}} \right), \qquad (5.141)$$

where $J_{fm} = 2t_{fm}t_{mf}/U$. Thus, H_{eff} (thereafter H_{t-J*}) can be written as

$$H_{t-J*} = H_{t-J} + H_{(3)},$$

$$H_{(3)} = \sum_{\substack{fmg\sigma \\ (f\neq g)}} \left(\frac{t_{fm}t_{mg}}{U}\right) \left(X_f^{\sigma 0} X_m^{\bar\sigma\sigma} X_g^{0\bar\sigma} - X_f^{\sigma 0} X_m^{\bar\sigma\bar\sigma} X_g^{0\sigma}\right). \quad (5.142)$$

Let us discuss the differences between H_{t-J*} for the metallic phase and for the case of half-filling. At $n = 1$, each site is occupied by one electron. Therefore, the Hilbert subspace for H_{t-J*} is formed by a set of homopolar states where both doubly occupied states and holes as well are absent. In this case, the projection operator can be written as

$$P = \prod_f \left(X_f^{\uparrow\uparrow} + X_f^{\downarrow\downarrow}\right),$$

and H_{t-J*} is reduced to the Heisenberg model. On the other hand, if $n < 1$, holes are present in the system. Therefore, electron hopping processes (the second term in Eq. (5.141)) and three-site interactions $H_{(3)}$ are included in H_{t-J*}.

It is evident from Eq. (5.142) that the parametric smallness of three-site interactions $H_{(3)}$ is the same as that of the exchange part of the t–J model, but it is considerably smaller than that of the kinetic part. This fact explains the negligible effect of $H_{(3)}$ on the spectral properties of the system in the normal phase. The situation is different when a superconducting state with d-type symmetry of order parameter (OP) is formed. In this case, the coupling constant in the superconducting phase $J \sim t^2/U$ is of the same order of magnitude as the three-site interactions. This is why one should expect that $H_{(3)}$ will strongly affect superconductivity with the d-type symmetry OP.

To study the superconductivity we introduce the normal Green's functions

$$D_{0\sigma,0\sigma}(f\tau, g\tau') = -\langle T_\tau \tilde{X}_f^{0\sigma}(\tau)\tilde{X}_g^{\sigma 0}(\tau')\rangle,$$
$$D_{\sigma 0,\sigma 0}(f\tau, g\tau') = -\langle T_\tau \tilde{X}_f^{\sigma 0}(\tau)\tilde{X}_g^{0\sigma}(\tau')\rangle \quad (5.143)$$

and anomalous Green's functions

$$D_{0\sigma,\bar\sigma 0}(f\tau, g\tau') = -\langle T_\tau \tilde{X}_f^{0\sigma}(\tau)\tilde{X}_g^{0\bar\sigma}(\tau')\rangle,$$
$$D_{\sigma 0,0\bar\sigma}(f\tau, g\tau') = -\langle T_\tau \tilde{X}_f^{\sigma 0}(\tau)\tilde{X}_g^{\bar\sigma 0}(\tau')\rangle. \quad (5.144)$$

As have been shown in Eq. (5.100), the Green's function $\hat{D}$ is expressed by the propagator function $\hat{G}$ and the strength operator $\hat{P}$.

$$D_{\alpha\beta}(\boldsymbol{k},\omega_n) = \sum_{\gamma} G_{\alpha\gamma}(\boldsymbol{k},\omega_n) P_{\gamma\beta}(\boldsymbol{k},\omega_n). \tag{5.145}$$

The Dyson equation for normal and anomalous propagators looks like

$$\tag{5.146}$$

Here the thick line corresponds to the normal propagator $G_{0\uparrow,0\uparrow}(\boldsymbol{p},\omega_n)$, and the thick line notes the anomalous propagator $G_{\downarrow 0,0\uparrow}(\boldsymbol{p},\omega_n)$. The Hartree–Fock propagator $G^{(0)}_{0\uparrow,0\uparrow}(\boldsymbol{p},\omega_n)$ is shown by the double thin line and is given by

$$\tag{5.147}$$

A solution of this equation reads

$$G^{(0)}_{0\uparrow,0\uparrow}(\boldsymbol{p},\omega_n) = (i\omega_n - \varepsilon + \mu - P_{11}t_{\boldsymbol{p}})^{-1}, \tag{5.148}$$

where $P_{11}(\boldsymbol{p},\omega_n) = P_{0\uparrow,0\uparrow}(\boldsymbol{p},\omega_n)$ is a component of the strength operator. The Green's functions (5.143), (5.144) have the following symmetry

$$\begin{aligned}
D_{\uparrow 0,\uparrow 0}(\boldsymbol{p},\omega_n) &= -D_{0\uparrow,0\uparrow}(-\boldsymbol{p},-\omega_n) = -D^*_{0\uparrow,0\uparrow}(-\boldsymbol{p},\omega_n), \\
D_{0\downarrow,\uparrow 0}(\boldsymbol{p},\omega_n) &= -D_{0\uparrow,\downarrow 0}(-\boldsymbol{p},-\omega_n), \\
D_{0\uparrow,\downarrow 0}(\boldsymbol{p},\omega_n) &= D^*_{\downarrow 0,0\uparrow}(\boldsymbol{p},-\omega_n).
\end{aligned} \tag{5.149}$$

This symmetry and Dyson equation result in the dispersion relations for the self-energy matrix

$$\Sigma_{22}(\boldsymbol{p},\omega_n) = -\Sigma_{11}(-\boldsymbol{p},-\omega_n), \qquad \Sigma_{21}(\boldsymbol{p},\omega_n) = \Sigma^*_{12}(\boldsymbol{p},-\omega_n). \tag{5.150}$$

In the standard notations

$$\Sigma(\boldsymbol{p},\omega_n) \equiv \Sigma_{11}(\boldsymbol{p},\omega_n), \qquad \Delta(\boldsymbol{p},\omega_n) \equiv \Sigma_{12}(\boldsymbol{p},\omega_n), \tag{5.151}$$

the diagram (5.146) has the form of the Gor'kov equations

$$[i\omega_n - \xi_{\bm{p}}(\omega_n)]G(\bm{p},\omega_n) - \Delta(\bm{p},\omega_n)F(\bm{p},\omega_n) = 1,$$
$$-\Delta^*(\bm{p},-\omega_n)G(\bm{p},\omega_n) + [i\omega_n + \xi_{-\bm{p}}(-\omega_n)]F(\bm{p},\omega_n) = 0.$$
(5.152)

Here we denote the normal function $G_{0\uparrow,0\uparrow}(\bm{p},\omega_n)$ by $G(\bm{p},\omega_n)$, and the anomalous one $G_{\downarrow 0,0\uparrow}(\bm{p},\omega_n)$ by $F(\bm{p},\omega_n)$. The renormalization of the normal state quasiparticle spectrum is given by

$$\xi_{\bm{p}}(\omega_n) = \varepsilon - \mu + P_{11}(\bm{p},\omega_n)t_{\bm{p}} + \Sigma_{11}(\bm{p},\omega_n). \tag{5.153}$$

Solution of the Gor'kov equations (5.152) reads

$$G(\bm{p},\omega_n) = \frac{i\omega_n + \xi_{-\bm{p}}(-\omega_n)}{[i\omega_n - \xi_{\bm{p}}(\omega_n)][i\omega_n + \xi_{-\bm{p}}(-\omega_n)] - \Delta(\bm{p},\omega_n)\Delta^*(\bm{p},-\omega_n)},$$

$$F(\bm{p},\omega_n) = \frac{\Delta^*(\bm{p},-\omega_n)}{[i\omega_n - \xi_{\bm{p}}(\omega_n)][i\omega_n + \xi_{-\bm{p}}(-\omega_n)] - \Delta(\bm{p},\omega_n)\Delta^*(\bm{p},-\omega_n)}.$$
(5.154)

To calculate the components of the self-energy and strength operators we introduce the following notations for the interaction processes. The electron hopping

$$t_{fm} X_f^{\sigma 0} X_m^{0\sigma} \tag{5.155}$$

is shown by wavy line $\overset{\displaystyle\sim\!\sim\!\sim\!\blacktriangleright}{\sigma 0 \quad\quad 0\sigma}$, while the exchange interaction

$$J_{fm} X_f^{\uparrow\downarrow} X_m^{\downarrow\uparrow}, \qquad J_{fm} X_f^{\sigma\sigma} X_m^{\bar{\sigma}\bar{\sigma}} \tag{5.156}$$

is denoted by dotted line

$$\underset{+-\quad\quad-+}{-\!-\!-\!-\!-\!-\!-\!-}\,, \qquad \underset{++\quad\quad--}{-\!-\!-\!-\!-\!-\!-\!-} \;. \tag{5.157}$$

The three-site hopping $H_{(3)}$ terms

$$\left(\frac{t_{fm}t_{mg}}{U}\right) X_f^{\sigma 0} X_m^{\bar{\sigma}\sigma} X_g^{0\bar{\sigma}}, \qquad \left(\frac{t_{fm}t_{mg}}{U}\right) X_f^{\sigma 0} X_m^{\bar{\sigma}\bar{\sigma}} X_g^{0\sigma}, \tag{5.158}$$

will be given by (here $f \neq g$)

$$\underset{f\quad\quad m\quad\quad g}{\overset{0\bar{\sigma}\quad\quad\bar{\sigma}\sigma\quad\quad\sigma 0}{\longleftarrow\!\sim\!\sim\!\blacktriangle\!\sim\!\sim\!\longrightarrow}}\,, \qquad \underset{f\quad\quad m\quad\quad g}{\overset{0\sigma\quad\quad\bar{\sigma}\bar{\sigma}\quad\quad\sigma 0}{\longleftarrow\!\sim\!\sim\!\circ\!\sim\!\sim\!\longrightarrow}} \;. \tag{5.159}$$

In the mean field approximation the strength operator $P_{11}(\boldsymbol{p},\omega_n)$ is given by a filling factor (bare circle) and equal to $P_{11} = (1 - n/2)$. In the same approximation a contribution to Σ_{11} results from three-site hopping and is given by the graph:

Analytically this contribution for $\Sigma_{11}(\boldsymbol{k})$, is equal to (neglecting the chemical potential renormalization)

$$\Sigma_{11}(\boldsymbol{k}) = -\frac{n}{2}\left(1 - \frac{n}{2}\right)\frac{t_{\boldsymbol{k}}^2}{U}.$$

Thus in the mean field approximation,

$$\xi_{\boldsymbol{p}}(\omega_n) = \varepsilon - \mu + (1 - n/2)t_{\boldsymbol{p}} - n/2(1 - n/2)t_{\boldsymbol{k}}^2/U. \tag{5.160}$$

For the anomalous component of the self-energy $\Delta(\boldsymbol{p},\omega_n)$, the electron hopping (5.155) results in the kinematic mechanism of pairing (Zaitsev and Ivanov 1987)

$$\tag{5.161}$$

An analytical expression for these two graphs is given by

$$\Delta_{(t)}(\boldsymbol{p},\omega_n) = -\frac{T}{N}\sum_{\boldsymbol{q}\omega_m} 2t_{\boldsymbol{q}}\, G_{0\uparrow,\downarrow 0}(\boldsymbol{q},\omega_m). \tag{5.162}$$

The exchange interaction J_{fm} in the H_{t-J} results in the magnetic mechanism of pairing and its contribution to $\Delta(\boldsymbol{p},\omega_n)$ is shown by the graphs

(Izyumov, Katsnelson and Skryabin 1994)

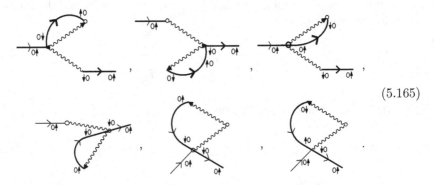

(5.163)

The corresponding contribution is equal to

$$\Delta_{(J)}(p,\omega_n) = -\frac{T}{N}\sum_{q\omega_m}(J_{p+q} + J_{p-q})G_{0\uparrow,\downarrow 0}(q,\omega_m). \qquad (5.164)$$

The three-site interactions produce 6 graphs for $\Delta_{(3)}(p,\omega_n)$

(5.165)

A sum of these graphs results in

$$\Delta_{(3)}(p,\omega_n) = -\frac{T}{N}\sum_{q\omega_m}\left[\left(\frac{4t_p t_q}{U} - J_{p+q} - J_{p-q}\right)(1-n/2)\right.$$
$$\left. - \left(\frac{t_q^2}{U} - \frac{1}{2}J_0\right)n\right]G_{0\uparrow,\downarrow 0}(q,\omega_m). \qquad (5.166)$$

The total anomalous self-energy is given by the summation of $\Delta_{(t)}$, $\Delta_{(J)}$ and $\Delta_{(3)}$, and is equal to

$$\Delta(p,\omega_n) = -\frac{T}{N}\sum_{q\omega_m}[2t_q + (J_{p+q} + J_{p-q}) + A_q^{(3)}]G_{0\uparrow,\downarrow 0}(q,\omega_m), \qquad (5.167)$$

where

$$A_q^{(3)} = \left(1-\frac{n}{2}\right)\frac{4t_p t_q}{U} - n\left(\frac{t_q^2}{U} - \frac{1}{2}J_0\right)\left(\frac{4t_p t_q}{U}\right)$$
$$- \left(1-\frac{n}{2}\right)(J_{p+q} + J_{p-q}). \qquad (5.168)$$

The last term in $A_q^{(3)}$ is the most significant factor for the d-symmetry OP. It leads in the final $\Delta(\boldsymbol{p},\omega_n)$ to the renormalization of the coupling constant by the scenario (Yushankhay, Vujicic and Zakula 1990) $J \to J[1 - (1 - n/2)] = (n/2)J$. It is this renormalization that determines the strong effect of $H_{(3)}$ on the formation of superconductivity with d symmetry OP. The self-consistent equation for OP can be found in the conventional way $(\Delta(\boldsymbol{p},\omega_n) \equiv \Delta_{\boldsymbol{p}})$:

$$\Delta_{\boldsymbol{p}} = \frac{1}{N}\sum_{\boldsymbol{q}}\left\{2t_{\boldsymbol{q}} + \frac{n}{2}(J_{\boldsymbol{p}+\boldsymbol{q}} + J_{\boldsymbol{p}-\boldsymbol{q}}) + 4\left(1 - \frac{n}{2}\right)\frac{t_{\boldsymbol{k}}t_{\boldsymbol{q}}}{U}\right.$$
$$\left. - n\left(\frac{t_{\boldsymbol{q}}^2}{U} - \frac{1}{2}J_0\right)\right\}\left(\frac{\Delta_{\boldsymbol{q}}}{2E_{\boldsymbol{q}}}\right)\tanh\left(\frac{E_{\boldsymbol{q}}}{2T}\right). \tag{5.169}$$

There are two distinctions of this equation from the corresponding equation for the t–J model. First, additional terms appear, which can be easily distinguished by the explicit dependence on the parameter U. The second distinction is more important and is associated with the renormalization of the coefficient before the terms $J_{\boldsymbol{p}\pm\boldsymbol{q}}$ indicated above.

The symmetry of the order parameter that is a solution of Eq. (5.169) may be different. The solution with the s-type symmetry $\Delta_{\boldsymbol{p}} = \Delta_0$ is not considered here for two reasons. First, it has been obtained previously by Zaitsev and Ivanov (1987), and the second reason is that the three-site hopping is cancelled for the s-symmetry solution. The solution with the d symmetry OP:

$$\Delta_{\boldsymbol{p}} = \Delta_0(\cos p_x - \cos p_y)$$

is of most interest. In this case, the equations for the temperature dependence $\Delta_0(T)$ and for calculating the critical temperature T_c can be written as follows:

$$1 = \frac{nJ}{2N}\sum_{\boldsymbol{q}}\frac{(\cos q_x - \cos q_x)^2}{E_{\boldsymbol{q}}}\tanh\left(\frac{E_{\boldsymbol{q}}}{2T}\right),$$
$$1 = \frac{nJ}{2N}\sum_{\boldsymbol{q}}\frac{(\cos q_x - \cos q_x)^2}{\varepsilon_{\boldsymbol{q}} - \mu}\tanh\left(\frac{\varepsilon_{\boldsymbol{q}} - \mu}{2T}\right). \tag{5.170}$$

The results of a numerical solution of the equation for T_c are given in Fig. 5.1 at various concentration n for the t–J^* model (curve 2). For comparison, the dependence of T_c on the electron concentration obtained by omitting $H_{(3)}$ is also given in this figure (curve 1). The numerical calculations were

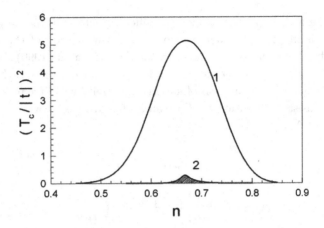

Fig. 5.1 Regions of the superconducting state in the (1) t–J and (2) t–J^* models.

performed under the assumption that the ratio $2|t|/U$ equals 0.25. It is evident that the inclusion of $H_{(3)}$ leads to a significant decrease in the superconducting transition temperature (hatched region).

The decrease in the critical temperature caused by $H_{(3)}$ is due to two factors. The first (and the main) factor is in the renormalization of the coupling constant. The second factor is manifested by the additional renormalization of the electron energy spectrum.

Figure 5.2 shows the dependence of T_c on the ratio $2|t|/U$ obtained without accounting for three-center interactions (curve 1) and with these interactions (curve 2). The electron concentration in the plots corresponds to the optimal doping level ($n = 0.665$). It is evident that, at $2|t|/U = 0.25$ (dashed lines), T_c obtained when considering the effect of $H_{(3)}$ is 25 times smaller than T_c calculated without the three-center interactions.

In order to clarify the physical nature of the renormalization of coupling constant by three-center interactions, consider the action of the H_{t-J} and $H_{(3)}$ operators on a singlet pair. If there are no other electrons, the state of a system with a given pair is described by the ket vector

$$|\Psi(f, f + \delta)\rangle = \frac{1}{\sqrt{2}}(X_f^{\uparrow 0} X_{f+\delta}^{\downarrow 0} - X_f^{\downarrow 0} X_{f+\delta}^{\uparrow 0})|0\rangle,$$

where $|0\rangle$ is the state without electrons. This pair corresponds to an eigenvector of H_{t-J}:

$$H_{t-J}|\Psi(f, f + \delta)\rangle = (2\varepsilon - 4t^2/U)|\Psi(f, f + \delta)\rangle.$$

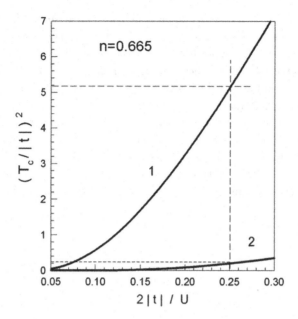

Fig. 5.2 Effect of $H(3)$ on the dependence of T_c on the parameter $2|t|/U$.

After the action of $H_{(3)}$ on the singlet pair, a superposition of states is obtained

$$H_{(3)}|\Psi(f,f+\delta)\rangle = (-4t^2/U) \sum_{\delta_1 \neq \delta} \{|\Psi(f,f+\delta_1)\rangle + |\Psi(f+\delta+\delta_1,f+\delta)\rangle\}.$$

It is evident that the effect of $H_{(3)}$ is reduced to rotations through the angles $\pi/2$, and $3\pi/2$ of the singlet pair under consideration around sites f and $f + \delta$. It is essential that the energy parameter in this case equals to $-2t^2/U$. As a result, when the right-hand side of the last equation is written in the form that does not contain the restriction $\delta_1 \neq \delta$ (one has to make a provision of this kind when passing over to the Fourier representation), the term $-4t^2/U|\Psi(f,f+\delta)\rangle$ should be added, and this term fully compensates the action of the exchange part of the Hamiltonian H_{t-J}. Thus, it can be seen that three-center interactions make a significant contribution to the dynamics of singlet pairs, whose formation lies in the basis of the mechanism of superconducting pairing. Therefore, in the case when the system contains only one singlet pair, one may comment on the full compensation of the corresponding two-center terms of the effective Hamiltonian. If the system contains other electrons, three-center terms act in such a way that the states arising because of changes in the lattice sites adjacent to the singlet pair under consideration start to make a contribution to the resulting superposition. These additional contributions increase as electrons fill sites adjacent to the pair. These circumstances explain the appearance of the concentration factor, which leads to the renormalization of the coupling constant mentioned above.

Note in conclusion that the analysis performed unambiguously points to an essential role of three-center interactions in the formation of a superconducting state with d-type symmetry OP. Whereas it was obtained without $H_{(3)}$ that $T_c \sim 100\,\text{K}$ at typical values of parameters, we obtained that accounts of these terms that $T_c \sim 4\,\text{K}$ for the same parameters.

Chapter 6

Spectral Properties of Anisotropic Metallic Ferromagnets

The effects of the crystal field on the spectral and thermodynamic properties of metallic magnetic compounds are well-known (Buschow 1979; Fulde and Loewenhaupt 1986; Kurtovoi and Levitin 1987). The rare-earth intermetallic compounds $ReAl_2$ (Re is a rare-earth element) and the actinide compounds are examples of such systems. The single-ion anistropy (SA) in these materials is of the same order of magnitude as the interatomic exchange interaction induced by itinerant electrons. This occurs, for example, in $HoAl_2$ (Fulde and Loewenhaupt 1986; Barbara et al 1977; Rhyne and Koon 1983; Ibarra et al 1985; Schelp et al 1985), in the uranium monochalcogenides (Kurtovoi and Levitin 1987), and in UGa_2 (Andreev et al 1978).

In the conventional approach to the intermetallic magnets the effective interatomic exchange interaction induced by itinerant electrons is calculated without taking into account anisotropy and the external field that are included afterwards (Fulde and Loewenhaupt 1986; Lindgaard 1986). Strictly speaking, the second order perturbation theory results in the effective Heisenberg exchange interaction only in the absence of the magnetic field and SA (Taylor and Darby 1972; Vonsovskii 1975). Therefore, it is not clear up to what values of SA it is possible to extend the conventional approach. It is evident that at large SA the effective exchange interaction may depend on the degree of anisotropy. It is not clear also whether the itinerant electron degree of freedom will be integrated, giving the effective interaction or they will give some contribution beyond the effective interaction.

The nontrivial electron contribution to the magnetic properties of the rare-earth metals has been demonstrated by Dzyaloshinskii (1965). He showed that the topological peculiarities of the Fermi surface results in the instability of the uniform ferromagnetic state and the formation of the helicoid structure. In ferromagnetic phase induced by large external magnetic field, the spin-wave dispersion is determined by both the Zeeman splitting and the changes in the effective exchange interaction.

For the strong SA we can also expect two channels of the spin excitation spectrum modification. Firstly, the SA results in the nonequidistant energy level scheme inside the ion. Secondly, the effective exchange interaction is expected to depend on the degree of anisotropy. So, the large SA may modify not only the activation energy (gap) of magnons but also the spin-wave stiffness.

The effective interatomic exchange interaction in the case of a strong crystal field with non-zero orbital moment has been considered by Irkhin (1966), Druzhinin and Irkhin (1966), and Irkhin et al (1968) (see the review by Irkhin (1988)). A non-Heisenberg contribution and large magnetocrystalline anisotropy have been found. The spectral properties of such system have not been discussed.

The spin-wave spectrum in the metallic ferromagnet can be studied without considering the effective Heisenberg interaction. For the isotropic $s-f$ exchange model it was demonstrated long ago by Vonsovskii and Izyumov (1962). For the colinear magnet with the strong SA it has been done by Podmarkov and Sandalov (1984).

In this chapter the theory of the anisotropic metallic magnets is described in the framework of the $s-d(f)$ exchange model with strong SA of an arbitrary symmetry and external magnetic field of an arbitrary orientation (Val'kov 1989). In the Hubbard X-operator representation the diagram technique is used and the dispersion equation for the magnetic excitation spectrum is derived. The solution of this equation is analyzed for two cases: that of the single axis and cubic SA anisotropy.

It is shown that at the large field H and strong SA the concept of the effective Heisenberg model becomes inadequate. At long wavelength of the spin excitations, a strong anomaly occurs that cannot be described by the effective interatomic exchange interaction. The condition of validity of the effective Heisenberg model is found. For the cubic SA the excitation spectrum is studied analytically for three directions of easy magnetisation axes and the phase diagram is obtained.

6.1 The Hamiltonian of the Anisotropic $s-f$ Magnet in Atomic Representation

We consider here the metallic magnet with LMM formed by localized d^n (f^n) electron configuration with the total magnetic moment $\hat{J}_f$. For simplicity only the lowest multiplet of the d^n (f^n) configuration is included. The exchange interaction of local magnetic moment (LMM) with the itinerant electron with spin $\hat{S}_f$ is given by the standard s–$d(s$–$f)$ exchange model. The crystal field effect on f ions will be treated by the Wigner–Ekkart theorem and the Stevense operators (Knox and Gold 1970; Zvezdin et al 1985). The model Hamiltonian can be written in the following form

$$H = \sum_{k\sigma}(\varepsilon_k - \mu)a^+_{k\sigma}a_{k\sigma} - \sum_f(2\mu_B \boldsymbol{H}_0 \boldsymbol{S}_f + g\mu_B \boldsymbol{H}_0 \hat{\boldsymbol{J}}_f)$$

$$+ \sum_f \sum_{nm} B_{nm} O_n^m(f) - 2\sum_{fg} A_{fg}(\boldsymbol{S}_f \boldsymbol{J}_g), \quad (6.1)$$

where the first term describes the noninteracting electrons, the second term corresponds to the Zeeman splitting, the third term is the SA of an arbitrary symmetry and the last term is the nonlocal s–d exchange interaction. Here, the index "0" at H indicates the components of the field in the initial coordinate system determined by the crystal axes. These components are equal to

$$H_0^x = H\sin\Theta_H \cos\varphi_H, \quad H_0^Y = H\sin\Theta_H \sin\varphi_H, \quad H_0^z = H\cos\Theta_H,$$
(6.2)

where Θ_H and φ_H are the corresponding angles in the spherical coordinates.

In the simplest form of SA, given by the sum of operators O_2^0 and O_4^0 and magnetic field oriented along the z-axis, the operator S^z commutes with the Hamiltonian. It allows one to use the S^z operator eigenvectors as the basis for the X-operator construction and to study the spectral properties of the model (Zaitsev 1975, 1976) by a rather straightforward generalization of the spin operator diagram technique applied to the isotropic s–f model by Izyumov et al (1974).

When the symmetry of the SA is more complex than that of the single axis or the magnetic field orientation induces the non-colinear magnetic order the operator S^z does not commute with the Hamiltonian. It results in the increasing number of diagrams in every order of perturbation theory when working with the X-operators constructed with the help of the S^z operator eigenvectors (Lutovinov and Reizer 1979; Vedyaev and Nikalaev 1982; Podmarkov and Sandalov 1984). The topological structure of the diagrams also becomes more complicated in this case.

It is shown below that it is more convenient to treat the SA of arbitrary symmetry using the method developed for the anisotropic Heisenberg model by Val'kov et al (1985), and to work in the matrix notations for the electron Green's functions similar to the Gor'kov's (1958) method in the theory of superconductivity (see also Vonsovskii et al 1977).

At first we choose the other coordinate system where the localized spin vector is oriented along the z-axis by the unitary transformation

$$H \to H' = UHU^+, \qquad (6.3)$$

with the operator U given by

$$U = \prod_f \exp\left(i\Theta T_f^y\right) \exp\left(i\varphi T_f^z\right), \quad \boldsymbol{T}_f = \boldsymbol{J}_f + \boldsymbol{S}_f.$$

Here, the angles Θ and φ determine the direction of the LMM vector $\langle \boldsymbol{J}_f \rangle$ in the initial coordinate system. The transformed Hamiltonian (6.3) coincides with the Hamiltonian (6.1) after the substitution $\boldsymbol{H}_0 \to \boldsymbol{H}$, where the vector $\boldsymbol{H}$ has the following components

$$\begin{aligned} H^x &= H[\cos\Theta \sin\Theta_H \cos(\varphi_H - \varphi) - \cos\Theta_H \sin\Theta], \\ H^y &= H\sin\Theta_H \sin(\varphi_H - \varphi), \\ H^z &= H[\sin\Theta \sin\Theta_H \cos(\varphi_H - \varphi) + \cos\Theta_H \cos\Theta]. \end{aligned} \qquad (6.4)$$

The same unitary transformation results in the transform of the $O_n^m(f)$ operators,

$$\widetilde{O}_n^m(f) = U O_n^m(f) U^+. \qquad (6.5)$$

The itinerant electron Hamiltonian including effects of the external magnetic field and itinerant electron spin polarization is nondiagonal with respect to the spin indexes. This results from the fact that the effective field for the itinerant electrons is pointing in another direction compared to the new axis O_z. Taking this axis as the quantization axis for the itinerant electron spins we can diagonalize it by the well-known Bogolubov (u,v)-transformation in the following way

$$\begin{aligned} a_{k\uparrow} &= u c_{k\uparrow} - v c_{k\downarrow}, \\ a_{k\downarrow} &= \overset{*}{v} c_{k\uparrow} + u c_{k\downarrow}. \end{aligned} \qquad (6.6)$$

The itinerant part of the Hamiltonian H_c can be written in the form

$$H_c = \sum_{k\sigma}(\varepsilon_k - \mu - \sigma\Delta) c_{k\sigma}^+ c_{k\sigma}, \qquad (6.7)$$

with the (u,v)-coefficients given by

$$u = \sqrt{\frac{1+x}{2}}, \quad v = \left(\frac{H^-}{H_\perp}\right)\sqrt{\frac{1-x}{2}}, \quad x = H_c/\sqrt{H_c^2 + H_\perp^2},$$

$$H_c = H_z + \left[\frac{A_0}{\mu_B}\right] R_L, \quad R_L \equiv \langle J_f^z \rangle, \quad A_0 = \sum_g A_{fg}, \quad (6.8)$$

$$H^\pm = H_x \pm iH_y, \quad H_\perp^2 = H_x^2 + H_y^2.$$

The gap Δ between the opposite spin subbands in the Hamiltonian (6.7) is equal to

$$\Delta = 2\mu_B \sqrt{H_c^2 + H_\perp^2}. \quad (6.9)$$

To take into account the SA we use the atomic representation. Let the single-ion Hamiltonian $H_0(f)$ have the eigenvectors $|\Psi_n(f)\rangle$ that satisfy the Schrodinger equation

$$H_0(f)|\Psi_n(f)\rangle = E_n |\Psi_n(f)\rangle. \quad (6.10)$$

Here the Hamiltonian $H_0(f)$ is given by

$$H_0(f) = \sum_{nm} B_{nm} \tilde{O}_n^m(f) - \bar{\boldsymbol{H}} \cdot \boldsymbol{J}_f, \quad (6.11)$$

where the effective field for the LMM may be written as

$$\bar{\boldsymbol{H}} = g\mu_B \boldsymbol{H} + 2A_0 \langle \boldsymbol{S} \rangle,$$

$$\langle \boldsymbol{S} \rangle = \frac{R_c}{\sqrt{H_\perp^2 + H_c^2}} (H_x, H_y, H_c), \quad R_c = \frac{1}{N} \sum_{k\sigma} \sigma \langle c_{k\sigma}^+ c_{k\sigma} \rangle. \quad (6.12)$$

Using the basis set $|\Psi_n(f)\rangle$ we construct the atomic Hubbard operators

$$X_f^{nm} \equiv |\Psi_n(f)\rangle\langle\Psi_m(f)|, \quad (6.13)$$

and write the operator $\hat{J}_f$ in the X-operator representation

$$J_f^+ = \sum_\lambda \gamma_\perp(\lambda) X_f^\lambda, \quad J_f^z = \sum_\lambda \gamma_\parallel(\lambda) X_f^\lambda. \quad (6.14)$$

To decrease the number of various diagrams it is convenient to use the matrix Green's function formulation similar to the Gor'kov's equation in the theory of superconductivity. The two-component Fermi operators are determined in the following way

$$c_f^+ = \left(c_{f\uparrow}^+, c_{f\downarrow}^+\right).$$

In the matrix notations the s–f exchange interaction model may be written as

$$H = 2A_0 N R_L R_c \frac{H_c}{\sqrt{H_c^2 + H_\perp^2}} + \sum_{k\sigma}(\varepsilon_{k\sigma} - \mu)c_{k\sigma}^+ c_{k\sigma}$$
$$+ \sum_f \sum_n E_n h_{fn} - \sum_{fg}\sum_\lambda \left\{c_f^+, \hat{\Gamma}_{fg}(\lambda) c_f\right\} X_g^\lambda, \quad (6.15)$$

with $\varepsilon_{k\sigma} = \varepsilon_k - \sigma\Delta$ and the matrix $\hat{\Gamma}_{fg}(\lambda)$ given by

$$\hat{\Gamma}_{fg}(\lambda) = A_{fg}\left[\tilde{\gamma}_\perp^*(-\lambda)\hat{\Gamma}_\perp + \gamma_\perp(\lambda)\hat{\Gamma}_\perp^+ + \gamma_{||}(\lambda)\hat{\Gamma}_{||}\right], \quad (6.16)$$

$$\hat{\Gamma}_\perp = \begin{pmatrix} u\overset{*}{v}, & u^2 \\ -\overset{*}{v}{}^2, & -\overset{*}{v}u \end{pmatrix}, \quad \hat{\Gamma}_{||} = \begin{pmatrix} u^2 - |v|^2, & -uv \\ -u\overset{*}{v}, & |v|^2 - u^2 \end{pmatrix}. \quad (6.17)$$

In colinear geometry the (u,v)-transformation (6.6) is not needed and the matrix $\Gamma(\lambda)$ has simpler expression

$$\hat{\Gamma}_{fg}(\lambda) = A_{fg}\begin{pmatrix} \gamma_{||}(\lambda), & \tilde{\gamma}_\perp^*(-\lambda) \\ \gamma_\perp(\lambda), & -\gamma_{||}(-\lambda) \end{pmatrix}. \quad (6.18)$$

6.2 The Green Function and Dispersion Equation

To study spin fluctuation dynamics we introduce the finite temperature Green's function of the Bose type in the usual manner

$$D_{\alpha\beta}(f\tau; g\tau') = -\langle T_\tau \tilde{X}_f^\alpha(\tau)\tilde{X}_g^{-\beta}(\tau')\rangle = \frac{T}{N}\sum_k \exp\{i\mathbf{k}\cdot(\mathbf{R}_f - \mathbf{R}_g)$$
$$- i\omega_n(\tau - \tau')\}D_{\alpha\beta}(k), \quad k \equiv (\mathbf{k}, \omega_n). \quad (6.19)$$

This Green's function will be calculated by perturbation theory over the small parameter $A_0 g_0 \ll 1$, where g_0 is the density of states at the Fermi energy. The diagram equation for the Green's function (6.19) may be written as

$$\Rightarrow\!=\!=\!\!\Rightarrow\!\Rightarrow\!+\!\Rightarrow\!\!\bigotimes\!\!\Rightarrow\!\!\Rightarrow \qquad (6.20)$$

where besides the usual self-energy term M we use the force part (Bar'yakhtar et al 1984; Garanin and Lutovinov 1984). The fat line

corresponds to the zeroth-order Green's function $D^{(0)}_{\alpha\beta}(k)$ determined by the solution of the equation

$$\Rightarrow\!=\!\Rightarrow\!+\!\Rightarrow\!\!\bigcirc\!\!\Rightarrow. \qquad (6.21)$$

Here the bare line corresponds to the local propagator $D_\alpha(\omega_n) = (i\omega_n + \boldsymbol{\alpha}\cdot\mathbf{E})^{-1}$ (see Chapter 5), and lines without arrows denote the matrix Green's functions of the itinerant electrons

$$\hat{G}(p) = \begin{pmatrix} G_\uparrow(p) & 0 \\ 0 & G_\downarrow(p) \end{pmatrix}, \qquad (6.22)$$

with

$$G_\sigma(p) = [i\omega_m - \varepsilon_\sigma(\boldsymbol{p}) + \mu]^{-1}, \quad \omega_m = (2m+1)\pi T.$$

The full dot in Eq. (6.21) corresponds to the bare vertex $\Gamma_{\boldsymbol{k}}(\alpha)$ given by the Fourier transform of the matrix (6.16):

$$\Gamma_{fg}(\lambda) = \frac{1}{N}\sum_{\boldsymbol{k}}\Gamma_{\boldsymbol{k}}(\lambda)\exp\{i\boldsymbol{k}\cdot(\boldsymbol{R}_f - \boldsymbol{R}_g)\}, \qquad (6.23)$$

while the filling factor $B(\alpha) = \langle X^{pp}\rangle - \langle X^{qq}\rangle$ is denoted by the empty dot. As is usual in the matrix notations the spur over the spin indexes should be calculated for every closed electron loop. The matrix notation allows one to reduce greatly the number of diagrams, thus in the conventional Fermi operator presentation one would have four different electron loops in the Eq. (6.21).

The analytical expression for Eq. (6.20) may be written as

$$D_{\alpha\beta}(k) = D^{(0)}_{\alpha\beta}(k)B(\beta) - \sum_{\beta_1\beta_2} D^{(0)}_{\alpha\beta_1} M_{\beta_1\beta_2} D_{\beta_2\beta}. \qquad (6.24)$$

The Green's function is determined by the self-energy M and force operator B according to the solution of Eq. (6.25):

$$D = [D^{(0)-1} + M]^{-1}B, \quad B_{\alpha\beta} \equiv \delta_{\alpha\beta}B(\beta). \qquad (6.25)$$

Equation (6.21) for the Green's function $D^{(0)}$ is written as

$$D^{(0)}_{\alpha\beta} = \delta_{\alpha\beta}D_\alpha(\omega_n) + D_\alpha(\omega_n)B(\alpha)\frac{T}{N}$$
$$\times \sum_{p\alpha_1} Sp\{G(p)\Gamma_{\boldsymbol{k}}(-\alpha)G(p+k)\Gamma_{\boldsymbol{k}}(\alpha_1)\}D^{(0)}_{\alpha_1\beta}(k). \qquad (6.26)$$

We introduce the polarization term for the itinerant electrons in the usual manner

$$\prod_{\sigma\sigma'}(k) = A_{\boldsymbol{k}}^2\left(\frac{T}{N}\right)\sum_p G_\sigma(p)G_{\sigma'}(p+k)$$
$$= \frac{A_{\boldsymbol{k}}^2}{N}\sum_p \frac{n_\sigma(\boldsymbol{p}) - n_{\sigma'}(\boldsymbol{p}+\boldsymbol{k})}{i\omega_n + \varepsilon_\sigma(\boldsymbol{p}) - \varepsilon_{\sigma'}(\boldsymbol{p}+\boldsymbol{k})}, \qquad (6.27)$$

with $n_\sigma(\boldsymbol{p}) = \left[\exp\left(\frac{\varepsilon_{p\sigma}-\mu}{T}\right) + 1\right]^{-1}$ being the Fermi–Dirac function. To solve Eq. (6.26), we denote by $X_\beta^{\sigma\sigma'}(k)$ the following linear combination

$$X_\beta^{\sigma'\sigma}(k) = \sum_\alpha \Gamma^{\sigma\sigma'}(\alpha)D_{\alpha\beta}(k), \quad \Gamma^{\sigma\sigma'}(\alpha) \equiv \Gamma_{\boldsymbol{k}}^{\sigma\sigma'}(\alpha)/A_{\boldsymbol{k}}. \qquad (6.28)$$

The function $D^{(0)}_{\alpha\beta}$ is determined by the $X_\beta^{\sigma\sigma'}(k)$ functions through

$$D^{(0)}_{\alpha\beta} = \delta_{\alpha\beta}D_\alpha(\omega_n) + D_\alpha(\omega_n)B(\alpha)\sum_{\sigma\sigma'}\Gamma^{\sigma\sigma'}(-\alpha)\prod_{\sigma\sigma'}(k)\cdot X_\beta^{\sigma'\sigma}(k), \qquad (6.29)$$

which is obtained from the solution of the system of four linear equations

$$X_\beta^{\sigma'_1\sigma_1} = \Gamma_{(\beta)}^{\sigma'_1\sigma_1}D_\beta + \sum_{\sigma'\sigma}L_{\sigma\sigma'}^{\sigma_1\sigma'_1}\prod_{\sigma\sigma'}X_\beta^{\sigma'\sigma}, \qquad (6.30)$$

$$L_{\sigma\sigma'}^{\sigma_1\sigma'_1} \equiv \sum_\alpha \Gamma^{\sigma_1\sigma'_1}(\alpha)\Gamma^{\sigma\sigma'}(-\alpha)D_\alpha(\omega_n)B(\alpha). \qquad (6.31)$$

Introducing the (4×4) matrix W with elements

$$W_{\sigma_1\sigma_1',\sigma\sigma'} \equiv L^{\sigma_1\sigma_1'}_{\sigma\sigma'}(\omega_n) \prod_{\sigma\sigma'}(\boldsymbol{k},\omega_n), \qquad (6.32)$$

we can write the solution of Eq. (6.30) in the following way

$$X^{\sigma\sigma'}_\beta = \sum_{\sigma_1\sigma_1'}\{(I-W)^{-1}\}_{\sigma\sigma',\sigma_1\sigma_1'}\Gamma^{\sigma_1\sigma_1'}_{(\beta)}D_\beta(\omega_n). \qquad (6.33)$$

Here I is the unit matrix

$$I_{\sigma\sigma',\sigma_1\sigma_1'} = \delta_{\sigma_1\sigma}\delta_{\sigma'\sigma_1'}.$$

Equations (6.29) and (6.33) determine the Green's function $D^{(0)}_{\alpha\beta}$. When the self-energy term is neglected the quasiparticle spectrum in the Hartree–Fock approximation is given by the dispersion equation

$$\Delta(\boldsymbol{k}, i\omega_n \to \omega - i\delta) = \det\|I - W\| = 0. \qquad (6.34)$$

The case of colinear geometry

When the external magnetic field is applied along the LMM, the magnetic order is colinear. This case is analyzed in detail below. The dispersion equation (6.34) may be written as

$$\Delta(\boldsymbol{k}\omega_n) = \begin{vmatrix} \Delta_\|(\boldsymbol{k},\omega_n) & \Delta_{12}(\boldsymbol{k},\omega_n) \\ \Delta_{21}(\boldsymbol{k},\omega_n) & \Delta_\perp(\boldsymbol{k},\omega_n) \end{vmatrix} = 0, \qquad (6.35)$$

with the matrix elements given by

$$\Delta_\|(\boldsymbol{k}\omega_n) = \begin{vmatrix} 1 - L^{zz}(\omega_n)\prod_{\uparrow\uparrow}(\boldsymbol{k},\omega_n) & L^{zz}(\omega_n)\prod_{\downarrow\downarrow}(\boldsymbol{k},\omega_n) \\ L^{zz}(\omega_n)\prod_{\uparrow\uparrow}(\boldsymbol{k},\omega_n) & 1 - L^{zz}(\omega_n)\prod_{\downarrow\downarrow}(\boldsymbol{k},\omega_n) \end{vmatrix},$$

$$\Delta_\perp(\boldsymbol{k}\omega_n) = \begin{vmatrix} 1 - L^{+-}(-\omega_n)\prod_{\downarrow\uparrow}(\boldsymbol{k},\omega_n) & -L^{--}(\omega_n)\prod_{\uparrow\downarrow}(\boldsymbol{k},\omega_n) \\ -L^{++}(\omega_n)\prod_{\downarrow\uparrow}(\boldsymbol{k},\omega_n) & 1 - L^{+-}(\omega_n)\prod_{\uparrow\downarrow}(\boldsymbol{k},\omega_n) \end{vmatrix},$$

$$\Delta_{12}(\boldsymbol{k}\omega_n) = \begin{vmatrix} -L^{z+}(\omega_n)\prod_{\downarrow\uparrow}(\boldsymbol{k},\omega_n) & -L^{z-}(\omega_n)\prod_{\uparrow\downarrow}(\boldsymbol{k},\omega_n) \\ L^{z+}(\omega_n)\prod_{\downarrow\uparrow}(\boldsymbol{k},\omega_n) & L^{z-}(\omega_n)\prod_{\uparrow\downarrow}(\boldsymbol{k},\omega_n) \end{vmatrix},$$

$$\Delta_{21}(\boldsymbol{k}\omega_n) = \begin{vmatrix} -L^{z-}(-\omega_n)\prod_{\uparrow\uparrow}(\boldsymbol{k},\omega_n) & L^{z-}(-\omega_n)\prod_{\downarrow\downarrow}(\boldsymbol{k},\omega_n) \\ -L^{z+}(-\omega_n)\prod_{\uparrow\uparrow}(\boldsymbol{k},\omega_n) & L^{z+}(-\omega_n)\prod_{\downarrow\downarrow}(\boldsymbol{k},\omega_n) \end{vmatrix}.$$

Here, the functions $L(\omega_n)$ are equal to

$$L^{zz}(\omega_n) = \sum_\alpha |\gamma_\|(\alpha)|^2 D_\alpha(\omega_n) B(\alpha),$$

$$L^{z+}(\omega_n) = \sum_\alpha \gamma_\|(\alpha)\gamma_\perp(-\alpha) D_\alpha(\omega_n) B(\alpha),$$

$$L^{z-}(\omega_n) = \sum_\alpha \gamma_\|(\alpha)\overset{*}{\gamma}_\perp(\alpha) D_\alpha(\omega_n) B(\alpha),$$

$$L^{++}(\omega_n) = \sum_\alpha \gamma_\perp(\alpha)\gamma_\perp(-\alpha) D_\alpha(\omega_n) B(\alpha), \quad (6.36)$$

$$L^{--}(\omega_n) = \sum_\alpha \overset{*}{\gamma}_\perp(\alpha)\overset{*}{\gamma}_\perp(-\alpha) D_\alpha(\omega_n) B(\alpha),$$

$$L^{+-}(\omega_n) = \sum_\alpha |\gamma_\perp(\alpha)|^2 D_\alpha(\omega_n) B(\alpha).$$

We restrict ourselves here to the single axis and cubic symmetry of the SA ferromagnetic metals.

6.3 The Ferromagnetic Metal with Single Axis Anisotropy

There are two colinear geometric situations when the magnetic field is parallel or perpendicular to the anisotropy axis. We consider both cases.

(a) The field H is perpendicular to the anisotropy axis

The O_z axis is the direction of the field H and magnetization M, and the O_x is the SA axis. The Hamiltonian of SA is given by

$$H_a(f) = 2D(S_f^x)^2.$$

The colinear case may be realized when $D > 0$ (an easy plane case) at any values of the magnetic field or when $D < 0$ (an easy axis case) at the strong magnetic field (see below).

In this geometry the solution of the single-ion [Eq. (6.10)] and the calculation of the representation parameters $\gamma(\alpha)$ [Eq. (6.14)] result in the conditions

$$\gamma_\perp(\alpha) \times \gamma_\|(\alpha) = 0, \quad L^{z\pm}(\omega_n) = 0.$$

The dispersion equation (6.35) has a block structure corresponding to the longitudinal and transverse spin waves.

The spectrum of the longitudinal waves is given by the solution of the following equation

$$1 - L^{zz}(\omega)\left[\prod_{\uparrow\uparrow}(\boldsymbol{k},\omega_n) + \prod_{\downarrow\downarrow}(\boldsymbol{k},\omega_n)\right] = 0. \quad (6.37)$$

In the long wavelength limit when $|\boldsymbol{k}| \leq k_f\omega/\mu$ (k_f and μ are the Fermi wavenumber and the chemical potential), and assuming a quadratic dispersion for the effective mass of the itinerant electron band we obtain the polarization operator to be equal to

$$\prod_{\sigma\sigma}(\boldsymbol{k},\omega_n) = \frac{A^2 n_\sigma}{m\omega^2}k^2, \quad n_\sigma = \frac{1}{N}\sum_p \left[\exp\left(\frac{\varepsilon_{p\sigma}-\mu}{T}\right) + 1\right]^{-1}. \quad (6.38)$$

We calculate the $L^{zz}(\omega)$ function at low temperatures $T \ll T_c$ when the population of the second and other excited single ion states is negligibly small and for the case of weak SA which satisfies the condition

$$D \ll \bar{H} = 2A\sigma + g\mu_B H.$$

Here σ is the itinerant electron magnetization

$$\sigma = (n_\uparrow - n_\downarrow)/2.$$

Then, in the second order in $(D/\bar{H})$, we find that the $L^{zz}(\omega)$ function is given by

$$L^{zz}(\omega) = \frac{2J(2J-1)\varepsilon_{31}}{\omega^2 - \varepsilon_{31}^2}\left(\frac{D}{\bar{H}}\right)^2, \quad (6.39)$$

with ε_{31} being equal to

$$\varepsilon_{31} = 2\bar{H} + 4(J-1)D.$$

Finally, the longitudinal spin-wave spectrum has the following form

$$\omega_n(\boldsymbol{k}) = \varepsilon_{31} + \frac{J(2J-1)}{6\pi^2}(3\pi^2 n)^{1/3}\left(\frac{A}{\bar{H}}\right)^2\left(\frac{D}{\bar{H}}\right)^2 \mu\Omega_0^{2/3}k^2, \quad (6.40)$$

where Ω_0 is an atomic volume, and $n = n_\uparrow + n_\downarrow$ is the itinerant electron concentration.

For the transverse waves the dispersion equation is

$$\left[1 - L^{+-}(\omega)\prod\nolimits_{\uparrow\downarrow}(\boldsymbol{k},\omega)\right]\left[1 - L^{+-}(-\omega)\prod\nolimits_{\downarrow\uparrow}(\boldsymbol{k},\omega)\right]$$
$$- L^{++}(\omega)L^{--}(\omega)\prod\nolimits_{\uparrow\downarrow}(\boldsymbol{k},\omega)\prod\nolimits_{\downarrow\uparrow}(\boldsymbol{k},\omega) = 0. \quad (6.41)$$

There are several excitations originating from the intra-ionic transitions between crystal field levels and dispersion due to the exchange interaction. Again, in the low temperature and low anisotropy limit, we obtain

$$L^{+-}(\omega) = \frac{2J}{\omega - \varepsilon_{21}} + J(2J-1)\frac{2(J-1)\varepsilon_{21} - \omega}{\omega^2 - \varepsilon_{21}^2}\left(\frac{D}{\bar{H}}\right)^2, \quad (6.42)$$

$$L^{++}(\omega)L^{--}(\omega) = \left(\frac{2J\varepsilon_{21}}{\omega^2 - \varepsilon_{21}^2}\right)^2\left(\frac{(2J-1)^2 D^2}{\bar{H}^2}\right), \quad (6.43)$$

where $\varepsilon_{nm} = E_n - E_m$ is the intra-ionic excitation energy. For the polarization operator we find that in the long wavelength limit it is equal to

$$\prod\nolimits_{\uparrow\downarrow}(\boldsymbol{k},\omega) = \frac{2A^2\sigma}{\omega - \Delta} + \frac{(3\pi^2 n)^{1/3}}{3\pi^2}\mu A^2 \frac{\omega - \Delta x^2}{(\omega - \Delta)^3}\Omega_0^{2/3}\boldsymbol{k}^2, \quad (6.44)$$

with $x = \Delta/4\mu, \Delta = 2\mu_B H + 2A\langle J^2\rangle$. The expression (6.44) is valid for a region of moment that satisfies the condition (k_F is the Fermi wavenumber):

$$\frac{|\omega - \Delta|}{2\mu} \gg \frac{|\boldsymbol{k}|}{k_F}. \quad (6.45)$$

The polarization operator $\prod_{\downarrow\uparrow}(\boldsymbol{k},\omega)$ can be written using Eq. (6.44) and the identity

$$\prod\nolimits_{\downarrow\uparrow}(\boldsymbol{k},\omega) = \prod\nolimits_{\uparrow\downarrow}(\boldsymbol{k},-\omega). \quad (6.46)$$

The condition of long wavelength in this case is given by

$$\frac{|\omega + \Delta|}{2\mu} \gg \frac{|\boldsymbol{k}|}{k_F}. \quad (6.47)$$

For the lowest spin-wave excitation $\omega \ll |\Delta|$ and with the both conditions (6.45) and (6.47) obeying

$$|\boldsymbol{k}| \ll \frac{|\Delta|}{2\mu}k_F$$

we obtain the spectrum in the following form

$$\omega(k) = \sqrt{\left(\tilde{g}\mu_B H + c\Omega_0^{3/2}k^2\right)\left(\tilde{g}\mu_B H + 2\widetilde{D} + c\Omega_0^{3/2}k^2\right) + \delta^2}, \quad (6.48)$$

with the renormalized g-factor given by

$$\tilde{g} = \frac{gJ + 2\sigma}{J + \sigma}. \quad (6.49)$$

It is worth noting that there is no renormalization of g-factor for $g = 2$, the same peculiarity of the $g = 2$ case was mentioned earlier for the isotropic s–f ferromagnets by Vonsovskii (1975). The parameter $\widetilde{D}$ is renormalized both due to quantum spin effect and conduction electrons

$$\widetilde{D} = \frac{J(2J-1)}{J+\sigma}D. \quad (6.50)$$

The spin-wave stiffness is equal to

$$c = \frac{(3\pi^2 n)^{1/3}}{24\pi^2}\left(\frac{J}{J+\sigma}\right)\left(\frac{A^2 J}{\mu}\right)\left[1 - \frac{2\omega_0^2}{AJ}\frac{(\mu/AJ)^2}{\tilde{g}\mu_B H + \widetilde{D}}\right], \quad (6.51)$$

and the gap in the spectrum is given by

$$\omega_0 \equiv \omega(k = 0) = \sqrt{\omega_{cl}^2(0) + \delta^2}, \quad (6.52)$$

where $\omega_{cl}(0)$ is a usual phenomenological gap for the single axis ferromagnet in the transverse field

$$\omega_{cl}(0) = \sqrt{\tilde{g}\mu_B H(\tilde{g}\mu_B H + 2\widetilde{D})}, \quad (6.53)$$

and the value δ is due to quantum effects

$$\delta^2 = \frac{2J(2J-1)}{J+\sigma}\left(\frac{\tilde{g}\mu_B \bar{H}}{\bar{H}}\right)D^2. \quad (6.54)$$

The similar quantum contributions to the spin-wave gap was predicted before in the theory of anisotropic Heisenberg magnets (Oguchi 1960; Kukharenko 1975).

For the easy axis case $D < 0$ it is evident from Eq. (6.53) that our consideration is valid for strong magnetic fields given by the condition

$$H > \frac{2J(2J-1)}{\tilde{g}\mu_B(J+\sigma)}\left[|D| - \frac{D^2}{\tilde{H}}\right]. \tag{6.55}$$

For the second spin-wave mode the gap at $k = 0$ is obtained from the solution of Eq. (6.41) and is equal to

$$\Omega = 2|A|(J+\sigma) + \frac{A}{|A|}\cdot\left[\frac{g\sigma+2J}{J+\sigma}\mu_B H + \frac{\sigma(2J-1)}{J+\sigma}D\right].$$

The analysis of the spectrum parameters has shown that the second term in Eq. (6.51) is absent at $T = 0$ due to the Goldstone theorem ($\omega_0 = 0$) and may be neglected when the following inequality takes effect

$$\frac{2\omega_0^2}{|A|J}\cdot\frac{(\mu/AJ)^2}{\tilde{g}\mu_B H + \widetilde{D}} \ll 1. \tag{6.56}$$

Then, the dependence of spin-wave spectrum of $s-f$ magnet on the parameters D, H and $\bm{k}$ coincides with the corresponding dependencies obtained in the framework of the Heisenberg model with the effective RKKY interaction and the forthcoming additional SA. We call such approach "RKKY + SA". Thus, the inequality (6.56) gives the condition under which the "RKKY + SA" model is valid for the description of the low temperature dynamics and thermodynamics of the SA metallic ferromagnet in the transverse geometry.

For $D > 0$ the condition (6.56) can be written in a simpler form

$$\frac{2\tilde{g}\mu_B H}{|A|J}\left(\frac{\mu}{AJ}\right) \ll 1. \tag{6.57}$$

In a weak magnetic field given by

$$H \ll H_0, \quad H_0 = \left(\frac{|A|J}{2\tilde{g}\mu_B}\right)\left(\frac{AJ}{\mu}\right)^2$$

the spin-wave stiffness is independent of H. At $H \sim H_0$ the character of the dispersion changes. It is more pronounced for $A > 0$, where the increase of the wavenumber results in the decreasing excitation energy and there appears a minimum in the $\omega(k)$ curve. The cause of these changes is due to a peculiar frequency dependence of the polarization operator (Val'kov and Val'kova 1985) which determines the effective exchange interaction.

To illustrate this anomaly in more detailed we consider the isotropic $s-f$ ferromagnet. The long wavelength spectrum of magnons is given by Eq. (6.48) with $D = 0$ and may be written as

$$\omega(\mathbf{k}) = \tilde{g}\mu_B H + \frac{(3\pi^2 n)^{1/3}}{24\pi^2}\left(\frac{J}{J+\sigma}\right)\left(\frac{A^2 J}{\mu}\right)$$

$$\times \left[1 - \frac{2\tilde{g}\mu_B H}{AJ}\left(\frac{\mu}{AJ}\right)^2\right]\Omega_0^{2/3}\mathbf{k}^2. \qquad (6.58)$$

At $g = 2$ and $H = 0$ the same expression was obtained by Izyumov and Vonsovskii (Vonsovskii 1975). Beyond the $k \to 0$ region the magnon spectrum is given by the solution of the following equation

$$\omega = \varepsilon_{21} + 2J\prod\nolimits_{\uparrow\downarrow}(\mathbf{k},\omega), \quad \varepsilon_{21} = \tilde{g}\mu_B H + 2A\sigma.$$

The numerical solution of this equation is shown in the Fig. 6.1.

Here, the parameters typical for the trivalent rare-earth metals have been used: $n = 3$, the atomic volume $18\,\text{cm}^3/\text{mol}$, $m = 3m_0$ (Taylor and Darby 1972; Vonsovskii 1975). The Fermi energy with these parameters is equal to $\mu = 2.63\,\text{eV}$. The rare-earth ion g-factor is included in the $s-f$ exchange parameter as usual by a substitution $A \to A = A_{sf}(g-1)$, the value A_{sf} is taken from a fit to the paramagnetic Curie temperature and is equal to $A_{sf} = 0.19\,\text{eV}$. With these parameters for the Gd ($J = 7/2, g = 2$) the magnon anomaly is expected at the magnetic field $H_0 = 183\,\text{T}$ that is

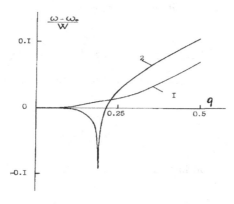

Fig. 6.1 The low lying magnon spectrum of the isotropic $s-f$-ferromagnetic metal at $H = 0$ (1) and $H > H_0$ (2).

a rather high value. For Tm^{3+} with $J = 6$, $g = 7/6$ we obtain $H_0 = 7.33\,\text{T}$ that may be achieved experimentally. In Fig. 6.1 the magnon spectrum is shown for the parameters corresponding to Tm and for magnetic field value equal to zero (curve 1) and $H = 10\,\text{T}$ (curve 2).

The values of the wavevector along the O_x axis are taken in units of $\sqrt{2m\Delta}$ with $k_B = \pi/a$ being equivalent to $q_B = 1.85$. The magnon energy is reckoned from $\omega_0 = \tilde{g}\mu_B H$ and measured in the magnon bandwidth W units. Here the bandwidth is determined in the following way

$$W = \omega(k = \pi/a), \quad a = \Omega_0^{1/3},$$

and is equal to $W = 5.07\,\text{K}$. At $H = 0$ there is a smooth anomaly of the magnon spectrum at $q = 0.5\sqrt{\Delta/\mu}$ that in units of k is given by

$$k \cong \Delta\sqrt{m/2\mu} \cong p_\uparrow - p_\downarrow,$$

with p_σ being the Fermi momentum of the electron with spin σ. Anomalous magnon spectrum can be studied experimentally in Tm which is a colinear ferromagnet above a magnetic field of $10\,\text{T}$. It is seen from Fig. 6.1 that at small q the magnon stiffness is negative and at $k = p_\uparrow - p_\downarrow$ a strong dip occurs. These anomalies are induced by the magnetic field and cannot be obtained from the "RKKY + SA" approach.

A similar behavior of the magnon spectra occurs at other model parameters. At the s–f exchange parameter $A < 0$ the spin-wave stiffness is positive for all values of H. Nevertheless, at $H \sim H_0$ it becomes dependent on the magnetic field as well as anisotropy.

(b) *Magnetic field H is parallel to the anisotropy axis*

The SA part of the Hamiltonian is written as an arbitrary analytical function of the S^z operator

$$H_a(f) = -\varphi\left(S_f^z\right). \qquad (6.59)$$

The dispersion equation has the following form

$$\left[1 - L^{+-}(\omega)\prod\nolimits_{\uparrow\downarrow}(\boldsymbol{k},\omega)\right]\left[1 - L^{+-}(-\omega)\prod\nolimits_{\downarrow\uparrow}(\boldsymbol{k},\omega)\right] = 0. \qquad (6.60)$$

At low temperatures and small wavenumber limit the magnon spectrum is given by

$$\omega(\mathbf{k}) = \omega_0 + \frac{(3\pi^2 n)^{1/3}}{24\pi^2} \left(\frac{J}{J+\sigma}\right) \left(\frac{A^2 J}{\mu}\right)$$
$$\times \left[1 - \left(\frac{2\omega_0}{AJ}\right) \left(\frac{\mu}{AJ}\right)^2\right] \Omega_0^{2/3} \mathbf{k}^2, \qquad (6.61)$$

with the gap in the spectrum equal to

$$\omega_0 = \tilde{g}\mu_B H + \left(\frac{J}{J+\sigma}\right)\varepsilon_a, \quad \varepsilon_a = \varphi(J) - \varphi(J-1). \qquad (6.62)$$

The expression for the gap is obtained using the inequality

$$|(g-2)\mu_B H + \varepsilon_a| \ll |A|J$$

that is quite realistic for s–f metal ferromagnets.

The gap for the upper magnon branch is given by

$$\Omega = 2|A|(J+\sigma) + \frac{A}{|A|}\left[\frac{g\sigma + 2J}{J+\sigma}\mu_B H + \frac{\sigma}{J+\sigma}\varepsilon_a\right]. \qquad (6.63)$$

Both gaps in Eqs. (6.62) and (6.63) reveal the itinerant electron renormalization of the effective anisotropy.

The magnetic field can influence the magnon spectrum. For example, in the simplest SA case when $\varphi(M) = DM^2$ the magnon gap is equal to

$$\omega_0 = \tilde{g}\mu_B H + \frac{J(2J-1)}{J+\sigma}D,$$

and for the easy axis magnets with $D > 0$ the value of the magnetic field H_0 that determines the anomalous itinerant electron contribution to the magnon spectrum is shifted to the low field region; at the large anisotropy it may occur even at $H = 0$. It means that the "RKKY + SA" approach cannot be used for

$$D \geq \left(\frac{2J-1}{2}\right)|A|J\left(\frac{AJ}{\mu}\right)^2$$

because in this approach the anisotropy just shifts the energy of magnon up without any anomalies in the spectrum.

6.4 The Metallic Ferromagnets with the SA of Cubic Symmetry

Some examples of the cubic crystal field symmetry are given by UX, X = S, Se, Te (Kurtovoi and Levitin 1987), and some ReAl$_2$ compounds (Buschow 1979; Fulde and Loewenhaupt 1986), PuSb (Lander $et\ al$ 1985). If the quantization axis is along one of the 4-th order axis of the crystal, then, using the Stevence operators (Knox and Gold 1964) the Hamiltonian of the SA can be written in the form (Buschow 1979; Fulde and Loewenhaupt 1986)

$$H_a(f) = \frac{Wx}{F_4} \left[O_4^0(f) + 5O_4^4(f)\right]$$
$$+ \frac{W(1-|x|)}{F_6} \left[O_6^0(f) - 21O_6^4(f)\right]. \quad (6.64)$$

Here the parameter $-1 < x < 1$ determines the relative contribution of the 4-th and 6-th order invariant to the anisotropy energy and the parameter W determines the level splitting. The coefficients F_4 and F_6 depend on the total angular momentum value J. The other expression for the cubic SA energy is closer to the phenomenological one

$$H_a = -\frac{1}{2}k_1\left(J_x^4 + J_y^4 + J_z^4\right) + \frac{1}{6}k_2\left(J_x^2 J_y^2 J_z^2 + J_x^6 + J_z^6 + J_y^6 + \cdots\right), \quad (6.65)$$

with the relations between parameters of the Eqs. (6.64) and (6.65) given by

$$\frac{Wx}{F_4} = -\frac{1}{40}\left[k_1 + \frac{3J(J+1) + 70}{33}k_2\right],$$
$$\frac{W(1-|x|)}{F_6} = \frac{1}{6 \cdot 11 \cdot 28}k_2. \quad (6.66)$$

We start with the case when the C_4 axis is an easy axis and is parallel to the magnetic field.

(a) H ||easy axis|| $[1,0,0]$

The single-ion Hamiltonian is given by Eq. (6.64) plus Zeeman term $-\bar{H}J^z$ with the effective field $\bar{H} = g\mu_B H + 2A\sigma$. The analysis of the single ion eigenvectors results in the following condition (Val'kov and Val'kova 1984)

$$\gamma_{||}(\alpha) \cdot \gamma_{\perp}(-\alpha) = 0 \qquad (6.67)$$

for all root vector α. This means the dispersion equations for the longitudinal and transverse waves are decoupled. The other peculiarity of the single ion eigenvectors is that we obtain

$$\gamma_{\perp}(\alpha) \cdot \gamma_{\perp}(-\alpha) = 0, \qquad (6.68)$$

for all α and from Eq. (6.36) one finds that

$$L^{++}(\omega) = L^{--}(\omega) = 0. \qquad (6.69)$$

The final form of the dispersion equation is the following:
for transverse mode

$$\left[1 - L^{+-}(\omega) \prod_{\uparrow\downarrow}(\boldsymbol{k},\omega)\right]\left[1 - L^{+-}(-\omega) \prod_{\downarrow\uparrow}(\boldsymbol{k},\omega)\right] = 0, \qquad (6.70)$$

for longitudinal mode

$$1 - L^{zz}(\omega)\left[\prod_{\uparrow\uparrow}(\boldsymbol{k},\omega) + \prod_{\downarrow\downarrow}(\boldsymbol{k},\omega)\right] = 0. \qquad (6.71)$$

The quantum effects due to SA are especially important at large anisotropy limit that requires the numerical solution of the Eqs. (6.70) and (6.71). When the anisotropy is weak relative to the magnetic field, the contribution of quantum renormalization may be written analytically. Thus, for the transverse mode at $T \ll T_c$, the function $L^{+-}(\omega)$ is given by

$$L^{+-}(\omega) = \frac{\gamma_{\perp}^2(1,2)}{\omega - \varepsilon_{21}} + \frac{\gamma_{\perp}^2(1,6)}{\omega - \varepsilon_{61}} - \frac{\gamma_{\perp}^2(4,1)}{\omega + \varepsilon_{41}}. \qquad (6.72)$$

This expression and Eq. (6.70) mean that the cubic SA results in the dispersion of the single ion excitations, $|\Psi_6\rangle \leftrightarrow |\Psi_1\rangle$ and $|\Psi_4\rangle \leftrightarrow |\Psi_1\rangle$, with the energies $\omega = \varepsilon_{61}$ and $\omega = \varepsilon_{41}$, and the magnon bandwidth is proportional to $W^2/\bar{H}$ which is small due to the inequality $W \ll \bar{H}$. The solution of Eq. (6.70) gives the gap in the optical magnon spectrum

$$\Omega = 2|A|(J+\sigma) + \frac{A}{|A|}\left[\frac{g\sigma+2J}{J+\sigma}\mu_B H + \left(\frac{\sigma}{J+\sigma}\right)\varepsilon_{100}\right], \quad (6.73)$$

where ε_{100} is the SA energy. In the notations of Eq. (6.64), this energy is equal to

$$\varepsilon_{100} = -\left(\frac{2k_J W}{F_4}\right)[10x + 21c_J(1-|x|)], \quad (6.74)$$

with the coefficient k_J given by

$$k_J = (2J-1)(J-1)(2J-3) \quad (6.75)$$

and caused by the kinematical renormalization of the anisotropy constant. It is clear that for $J = 1/2, 1, 3/2$ there is no cubic anisotropy. The parameter C_J in Eq. (6.74) is equal to

$$c_J = (J-2)(2J-5)F_4/F_6 \quad (6.76)$$

and becomes zero for $J = 2, 5/2$ due to the absence of the 6-th order invariant.

The energy of the lowest magnon transverse mode is obtained in a similar way. In the $k \to 0$ limit, it is given by

$$\omega_\perp = \omega_0 + c\Omega_0^{2/3} k^2, \quad (6.77)$$

with the spin-wave stiffness c equal to

$$c = \frac{(3\pi^2 n)^{1/3}}{24\pi^2}\left(\frac{J}{J+\sigma}\right)\left(\frac{A^2 J}{\mu}\right)\left[1 - \left(\frac{2\omega_0}{AJ}\right)\left(\frac{\mu}{AJ}\right)^2\right]. \quad (6.78)$$

The gap ω_0 can be written in the following way

$$\omega_0 = \tilde{g}\mu_B H + \left(\frac{J}{J+\sigma}\right)\varepsilon_{100} + \Delta\omega_Q, \quad (6.79)$$

with the effective g-factor $\tilde{g}$ given by Eq. (6.49). Here, the first two terms correspond to the phenomenological theory while the last one results from the quantum effects and is given by

$$\Delta\omega_Q = \left(\frac{J}{J+\sigma}\right)\left(\frac{k_J}{\bar{H}}\right)\left(\frac{10W}{F_4}\right)^2$$
$$\times \left\{Jf^2(x) - 3(J-2)[2Y^2(x) + f(x)Y(x)]\right\}, \qquad (6.80)$$

$$f(x) = -x - 21c_J(1-|x|), \quad Y(x) = f(x) + \frac{231}{5}\left(\frac{c_J}{J-2}\right)(1-|x|). \qquad (6.81)$$

The increasing W value increases also the quantum renormalization of the gap. At large W when $|\Delta\omega_Q| \geq \omega_0$, Eq. (6.79) is not valid and the numerical solution of the dispersion equation should be used.

The analysis of Eq. (6.78) reveals the anomaly in the magnon stiffness when the following inequality holds

$$\frac{2\omega_0}{|AJ|}\left(\frac{\mu}{AJ}\right)^2 \geq 1. \qquad (6.82)$$

For the longitudinal spectrum one can find

$$L^{zz}(\omega) = \left(\frac{48Jk_J}{\bar{H}}\right)\left(\frac{10W}{F_4}\right)^2 \frac{f^2(x)}{\omega^2 - \varepsilon_{51}^2}, \qquad (6.83)$$

and the magnon energy may be written in the form

$$\omega_\|(\mathbf{k}) = \varepsilon_{51} + Jk_J\frac{(3\pi^2 n)^{1/3}}{\pi^2}\left[\frac{5WAf(x)}{F_4\cdot\bar{H}^2}\right]^2 \mu\Omega_0^{2/3}\mathbf{k}^2,$$

$$\varepsilon_{51} = 4\bar{H} + \frac{W}{F_4}\left[x\left\langle 5\left|\hat{O}_4^0\right|5\right\rangle - \frac{F_4}{F_6}(1-|x|)\left\langle 5\left|\hat{O}_6^0\right|5\right\rangle\right] \qquad (6.84)$$

$$- \left(\frac{2Wk_J}{F_4}\right)[Jx - c_J(1-|x|)].$$

(b) $H\,\|$easy axis$\|\,[111]$

The O_z axis is chosen along C_3. The single ion Hamiltonian has the form

$$H_0(f) = -\frac{2Wx}{3F_4}\left(\hat{O}_4^0 + 20\sqrt{2}\hat{O}_4^3\right) + \frac{W(1-|x|)}{9F_6}$$
$$\times\left[16\hat{O}_6^0 - 140\sqrt{2}\hat{O}_6^3 + 154\hat{O}_6^6\right] - \bar{H}J^z. \qquad (6.85)$$

Similar to the previous case, one finds that

$$\gamma_{\|}(\alpha) \cdot \gamma_\perp(-\alpha) = 0, \quad \gamma_\perp(\alpha) \cdot \gamma_\perp(-\alpha) = 0.$$

The lowest branch of the transverse mode is given by

$$\omega_\perp = \omega_0 + c\Omega_0^{2/3} k^2, \tag{6.86}$$

where the gap is equal to

$$\omega_0 = \tilde{g}\mu_B H + \left(\frac{J}{J+\sigma}\right)\left(\frac{8Wk_J}{3F_4}\right)[5x - 28c_J(1-|x|)] + \Delta\omega_q.$$

The quantum renormalization of the gap has the following form

$$\Delta\omega_q = -\left(\frac{J}{J+\sigma}\right)\left(\frac{2k_J}{\bar{H}}\right)\left(\frac{40W}{3F_4}\right)^2 \left\{ \frac{(J-3)(18J-35)}{4}x^2 \right.$$
$$+ 7\left(6J^2 - \frac{133}{3}J + 68\right)c_J x(1-|x|)$$
$$\left. - 49\left[\left(\frac{11}{5}\right)^2 \frac{2(10-3J)F_4}{F_6} - \left(2J^2 - \frac{59}{3}J + \frac{175}{4}\right)c_J\right] c_J x(1-|x|)^2 \right\}, \tag{6.87}$$

and the magnon stiffness is given by Eq. (6.78).

The gap for the optical magnon branch is obtained from Eq. (6.73) by substitution

$$\varepsilon_{100} \to \varepsilon_{111} = \left(\frac{8Wk_J}{3F_4}\right)[5x - 28c_J(1-|x|)].$$

For the longitudinal waves the $L^{zz}(\omega)$ function has the next form

$$L^{zz}(\omega) = \frac{2\gamma_{\|}^2(1,4)\varepsilon_{41}}{\omega^2 - \varepsilon_{41}^2} + \frac{2\gamma_{\|}^2(1,7)\varepsilon_{71}}{\omega^2 - \varepsilon_{71}^2}, \tag{6.88}$$

and there appear two magnon branches. The lowest one has the following dispersion relation

$$\omega_{\|}^{(1)}(\mathbf{k}) = \varepsilon_{41} + \frac{8(3\pi^2 n)^{1/3}}{\pi^2}J(2J-3)k_J \left(\frac{20WA}{9\bar{H}^2 F_4}\right)^2$$
$$\times \left[x + \frac{14}{3}c_J(1-|x|)\right]^2 \mu\Omega_0^{2/3} k^2, \tag{6.89}$$

while the upper magnon energy is given by

$$\omega_{\|}^{(2)}(\boldsymbol{k}) = \varepsilon_{71} + \frac{5(3\pi^2 n)^{1/3}}{3\pi^2} J(J-2)(2J-5)k_J$$
$$\times \left[\frac{8 \cdot 77 \cdot AW(1-|x|)}{9F_6 \bar{H}^2}\right]^2 \mu\Omega_0^{2/3} \boldsymbol{k}^2. \quad (6.90)$$

(c) $\boldsymbol{H} \,\|\text{easy axis}\| \,[110]$

Taking the quantization axis along C_2, we can write the single ion Hamiltonian in the form

$$H_0 = \frac{Wx}{4F_4}\left(-\hat{O}_4^0 - 20 \cdot \hat{O}_4^2 + 15 \cdot \hat{O}_4^4\right)$$
$$+ \frac{W(1-|x|)}{8F_6}\left[-13 \cdot \hat{O}_6^0 + \frac{105}{2}\hat{O}_6^2 + 105 \cdot \hat{O}_6^4 + \frac{231}{2}\hat{O}_6^6\right] - \bar{H}J^z. \quad (6.91)$$

Contrary to two previous cases, here $L^{++}(\omega) \neq 0$ and $L^{--}(\omega) \neq 0$. Neglecting small quantum renormalizations we can write the spectrum of the lowest transverse mode in the following way:

$$\omega_\perp^2 = \left\{\tilde{g}\mu_B H + \left(\frac{J}{J+\sigma}\right)\left(\frac{2Wk_J}{F_4}\right)[10x + 21c_J(1-|x|)] + c\Omega_0^{2/3}\boldsymbol{k}^2\right\}$$
$$\times \left\{\tilde{g}\mu_B H + \left(\frac{J}{J+\sigma}\right)\left(\frac{2Wk_J}{F_4}\right)\left[-10x + \frac{189}{2}c_J(1-|x|)\right]\right.$$
$$\left. + c\Omega_0^{2/3}\boldsymbol{k}^2\right\}, \quad (6.92)$$

with the parameters given by

$$c = \left(\frac{J}{J+\sigma}\right)\frac{(3\pi^2 n)^{1/3}}{24\pi^2}\left(\frac{A^2 J}{\mu}\right)\left[1 - \left(\frac{2\omega_0^2}{AJ}\right)\frac{(\mu/AJ)^2}{\tilde{g}\mu_B H + \tilde{\varepsilon}_{110}}\right],$$
$$\tilde{\varepsilon}_{110} = \left(\frac{J}{J+\sigma}\right)\left(\frac{Wk_J}{F_4}\right)\left[5x + \frac{21 \cdot 13}{4}c_J(1-|x|)\right], \quad \omega_0 = \omega(\boldsymbol{k}=0).$$
$$(6.93)$$

In the absence of the magnetic field the region of ferromagnetic phase with the magnetization oriented along the C_2 axis is restricted by the following inequalities

$$W > 0, \quad -\frac{21c_J}{10+21c_J} < x < \frac{189c_J}{20+189c_J}. \tag{6.94}$$

For $J = 2, 5/2$ the value c_J is equal to zero and ferromagnetic order is impossible. The increasing magnetic field and anisotropy result in the anomaly in the spectrum, similar to that of Fig. 6.1.

The longitudinal oscillations have three branches given by

$$\omega_{\parallel}^{(1)}(\mathbf{k}) = \varepsilon_{31} + \frac{3}{2}J(J-1)(2J-3)k_J(3\pi^2 n)^{1/3} \left(\frac{5AW}{\pi F_4 \bar{H}}\right)^2$$
$$\times \left[x - \frac{7}{4}c_J(1-|x|)\right]^2 \mu \Omega_0^{2/3} \mathbf{k}^2,$$

$$\omega_{\parallel}^{(2)}(\mathbf{k}) = \varepsilon_{51} + Jk_J(3\pi^2 n)^{1/3}$$
$$\times \left(\frac{15AW}{4\pi F_4 \bar{\bar{H}}^2}\right)^2 \left[x + \frac{35}{2}c_J(1-|x|)\right]^2 \mu \Omega_0^{2/3} \mathbf{k}^2,$$

$$\omega_{\parallel}^{(3)}(\mathbf{k}) = \varepsilon_{71} + 15J(J-2)(2J-5)k_J(3\pi^2 n)^{1/3}$$
$$\times \left[\frac{77AW(1-|x|)}{4\pi F_6 \bar{\bar{H}}^2}\right]^2 \mu \Omega_0^{2/3} \mathbf{k}^2. \tag{6.95}$$

Comparing the ω_0 values for the considered geometries we can find the following phase diagram (Fig. 6.2) where the regions of stability for each kind of ordering are shown. This diagram is valid only in the weak anisotropy limit because the quantum renormalization was not taken into

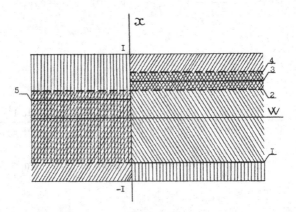

Fig. 6.2 The phase diagram of the cubic $s-f$-ferromagnets with the different orientations of the magnetization. ([1,0,0] (|||||), [1,1,1] (/////), [1,1,0] (\\\\\)).

account. The characteristic lines in Fig. 6.2 are given by

(1) $x = -21c_J/(10 + 21c_J)$;
(2) $x = 28c_J/(5 + 28c_J)$;
(3) $x = 49c_J/(6 + 49c_J)$;
(4) $x = 189c_J/(20 + 189c_J)$;
(5) $x = 7c_J/(15 + 7c_J)$;

The coefficient c_J determines the dependence of phase diagram on the J value. Thus, for $J = 2, 5/2$ the coefficient c_J is equal to zero and all lines coincide with the line $x = 0$. The phase M_{110} becomes impossible and the transition between the phases M_{100} and M_{111} with increasing anisotropy is a second order transition with the excitation energy equal to zero at the transition point. For $J > 5/2$, the phase M_{110} occurs at the $W > 0$ region. At the transition from M_{110} in M_{100} (line 1), the magnon energy is equal to zero and the spectrum becomes linear in the wavenumber: this is a second order transition. If we increase x starting from the M_{110} phase then above line 2 the system is in the region where both M_{110} and M_{111} phases are stable. The comparison of their energies has shown that below the line 3 the M_{110} phase has lower energy while above the line 3 the M_{111} phase has lower energy with M_{110} being the metastable phase up to the line 4. The phase transition $M_{110} \leftrightarrow M_{111}$ is of the first order with hysteresis between lines 2 and 4 (Landau and Lifshits 1980). The magnon frequency is not equal to zero at the line 3 while $\omega_{110} = 0$ at the line 4 and $\omega_{111} = 0$ at the line 2.

In the region $W < 0$, the phase M_{111} is metastable above the line 5 and the phase M_{100} is metastable below the line 5.

Chapter 7

Peculiarities of the de Haas-van Alphen Effect in Strongly Correlated Systems with Magnetic Polaron States

The de Haas-van Alphen (dHvA) effect provides the direct experimental study of the electronic structure and its role is especially important in the SEC systems because the theoretical methods of the band structure calculations for the SEC systems are not so developed as the conventional band theory. The experimental measurements of the dHvA oscillations in the heavy fermion compounds $CeCu_6$ (Reinders et al 1986), UPt_3 (Taillefer and Lonzarich 1988), $CeCu_2Si_2$ (Hunt et al 1990) have proved the existence of the Fermi surface and low damping Fermi-type quasiparticles near the Fermi level and have revealed some peculiarities of the dHvA effect in comparison with the conventional Fermi liquid materials.

For example, in $CeCu_2Si_2$ there is a spin-flip transition at the magnetic field $H = H_c$, the dHvA oscillations frequency has an abrupt change at $H = H_c$ and the amplitude of the oscillations near H_c has an anomalous field dependence. The nonlinear contribution to the dHvA oscillation due to magnetic order have been studied previously by Russul and Schlottmann (1990), and Sollie and Schlottmann (1990) however the changes of the period at the spin-flip transition have not been obtained. The theory of the dHvA effects has been considered in the framework of the periodic Anderson model (Wassermann et al 1989; Rasul 1989, the review paper by Wassermann and Springford 1996).

Recently, the dHvA effect has been studied in the SEC systems with low carrier concentration like RX where R is a rare-earth element and X = Bi, Sb, As, P. The long magnetic order and magnetopolaron states are essential for understanding the peculiar transport and thermodynamic properties of these materials (Kasuya et al 1993; Kasuya 1995). For CeX compounds the carrier concentration per R ion is very small, $n \sim 10^{-2}-10^{-3}$. The low values of the Neel temperature T_N (7 K for CeAs and 10.5 K for CeP) allows the experimental study of the magnetic field region where the dHvA oscillations occur in the vicinity of the spin-flip transition.

In the cubic NaCl crystal structure of CeX the bottom of the conductivity band is at the X point and the top of the valence band is at the Γ point of the Brillouin zone. A weak overlapping of these bands results in the semimetals properties. The valence band electrons have strong correlations with the localized f-electrons due to p–f hybridization. For homeopolar configurations of the Ce ions p–f hybridization results in the exchange interaction (Cogblin and Schrieffer 1969; Vonsovskii 1975; Nagaev 1988) between the localized and valence band electrons and it will be modelled below by isotopic s–f exchange interaction. As concerns the electrons of the conductivity band, the symmetry consideration (Kasuya et al 1993) gives the conclusion that their interaction with f electrons is small and we neglect it.

The strong spin correlations of p and f electrons may result in magnetic polaron states (Izyumov and Medvedev 1970). The valence bandwidth is small in comparison with the s–f interaction value and the magnetopolaron states in such narrow band antiferromagnets have been considered by Nagaev (1970, 1971) who has obtained the magnetopolaron band narrowing effect depending on the external magnetic field.

The magnetopolaron effect results in a large shift of the chemical potential of the semimetal that is important for the field dependence of the upper Landau levels. It is shown in this chapter that in a region of small carrier's concentration $n < n_c$ the number of the Landau levels increases at $H < H_c$ and decreases at $H > H_c$ (Val'kov and Dzebisashvili 1997a, 1997b). This results in the change of the dHvA oscillation period as was observed in the experimental data (Hunt et al 1990). This effect is of second order in magnitude in the small parameter $\theta \sim \sqrt{(H_c - H)/H_c}$, namely the angle between the sublattice magnetization and the external magnetic field. To calculate all θ^2 contributions, it is necessary to go beyond the quasiclassical approach and it is done here with the help of the X-operator representation.

One more example of the peculiarity of the dHvA effect has been found in the degenerate ferromagnetic semiconductors ACr_2Se_4 with A = Hg, Cd (Ovchinnikov et al 1995) where the quantum oscillations was measured at fixed magnetic field as a function of temperature. The origin of such anomalous dHvA oscillations is also the non-Fermi liquid contribution to the chemical potential which was induced by magnetic ordering.

7.1 The Hamiltonian of a Strong Correlated Narrow Band Antiferromagnet

Here we consider the s–f antiferromagnet with one band where the carriers are the holes at the top of the valence p-band. In the quasiclassical approach the narrow band limit of the model has been described in detail by Nagaev (1983). The X-operator formulation of the s–f model Hamiltonian was given in the Chapter 3. In an external magnetic field there appears a non-colinear geometry and to obtain the exact solution of the single ion eigenproblem we use the unitary transformation method. The other unusual property of the non-colinear case is the spin nondiagonal hopping of the electrons. Therefore, some generalization of the X-operator representation for the non-colinear s–f magnet will be presented here.

The model Hamiltonian has the following form

$$H = H_h + H_{s-d} + H_m, \qquad (7.1)$$

where the first term corresponds to the holes. In the antiferromagnet with two sublattices F and G, the hole Hamiltonian is written as

$$H_h = \sum_{ff'\sigma} [t_{ff'} - \delta_{ff'}(2\mu_B H\sigma - \mu)] c^+_{f\sigma} c_{f'\sigma} + \sum_{fg\sigma} t_{fg} \left(c^+_{f\sigma} d_{g\sigma} + \text{h.c.} \right)$$
$$+ \sum_{gg'\sigma} [t_{gg'} - \delta_{gg'}(2\mu_B H\sigma - \mu)] d^+_{g\sigma} d_{g'\sigma} + \sum_f U n_{f\uparrow} n_{f\downarrow} + \sum_g U n_{g\uparrow} n_{g\downarrow}$$
$$(7.2)$$

with $f \in F$, $g \in G$, and $c^+_{f\sigma}$ and $d^+_{g\sigma}$ the hole creation operators in the Wannier representation in sublattices F and G. The second term in Eq. (7.1) is the s–f-exchange interaction term in the two-sublattice form

$$H_{sd} = -A \sum_f (\boldsymbol{S}_f \boldsymbol{\sigma}_f) - A \sum_g (\boldsymbol{S}_g \boldsymbol{\sigma}_g), \qquad (7.3)$$

where $\boldsymbol{S}_{f(g)}$ and $\boldsymbol{\sigma}_{f(g)}$ are the spin operators of the localized and itinerant electrons. The last term in Eq. (7.1) describes the interatomic spin–spin interaction of the LMM and Zeeman energy

$$H_m = -\frac{1}{2}\sum_{ff'} I_{ff'} (S_f S_{f'}) - \frac{1}{2}\sum_{gg'} I_{gg'} (S_g S_{g'})$$
$$+ \sum_{fg} K_{fg} (S_f S_g) - g\mu_B H \left(\sum_f S_f^z + \sum_g S_g^z \right), \quad (7.4)$$

where I and K are the intra sublattice and intersublattice interactions correspondingly.

7.2 The Hamiltonian in the Non-colinear Phase: Unitary Transformation

In an external field the sublattices of the isotropic antiferromagnet are canted (Borovik-Romanov 1962; Gurevich 1973). To study the electron spectrum in the canted phase we use the local rotation of the coordinate system around the Oy axis through the angle θ for the F sublattice and the angle-θ for the G sublattice to direct the local magnetization vector along the new Oz axis. This rotation is described by the unitary transformation

$$H \to H' = \hat{U} H \hat{U}^+, \quad (7.5)$$

with the operator U given by

$$\hat{U} = \prod_f \prod_g \exp\left\{i\theta \left(S_f^y + \sigma_f^y\right)\right\} \exp\left\{-i\theta \left(S_g^y + \sigma_g^y\right)\right\}. \quad (7.6)$$

The electron operators transform in the following way (here $\bar{\sigma} \equiv -\sigma$)

$$c_{f\sigma}(\theta) = \hat{U} c_{f\sigma} \hat{U}^+ = c_{f\sigma} \cos\frac{\theta}{2} - 2\sigma c_{f\bar{\sigma}} \sin\frac{\theta}{2},$$
$$d_{g\sigma}(\theta) = \hat{U} d_{g\sigma} \hat{U}^+ = d_{g\sigma} \cos\frac{\theta}{2} + 2\sigma d_{g\bar{\sigma}} \sin\frac{\theta}{2}, \quad (7.7)$$

while the transformation of the spin operators is given by

$$S_f^x(\theta) \equiv \hat{U} S_f^x \hat{U}^+ = S_f^x \cos\theta + S_f^z \sin\theta,$$
$$S_f^y(\theta) \equiv \hat{U} S_f^y \hat{U}^+ = S_f^y, \quad (7.8)$$
$$S_f^z(\theta) \equiv \hat{U} S_f^z \hat{U}^+ = S_f^z \cos\theta - S_f^x \sin\theta.$$

The transformed Hamiltonian (7.5) may be written in the form

$$H'_h = \sum_{ff'\sigma} t_{ff'} c^+_{f\sigma} c_{f'\sigma} + \sum_{gg'\sigma} t_{gg'} d^+_{g\sigma} d_{g'\sigma}$$
$$+ \sum_{fg\sigma} t_{fg} (\cos\theta c^+_{f\sigma} d_{g\sigma} + 2\sigma \sin\theta c^+_{f\sigma} d_{g\bar{\sigma}} + \text{h.c.})$$
$$+ \sum_{f\sigma} \{\mu_B H \sin\theta c^+_{f\bar{\sigma}} c_{f\sigma} - (2\mu_B H \cos\theta \cdot \sigma - \mu) c^+_{f\sigma} c_{f\sigma}\} + \sum_f U n_{f\uparrow} n_{f\downarrow}$$
$$- \sum_{g\sigma} \{\mu_B H \sin\theta d^+_{g\bar{\sigma}} d_{g\sigma} + (2\mu_B H \cos\theta \cdot \sigma - \mu) d^+_{g\sigma} d_{g\sigma}\} + \sum_g U n_{g\uparrow} n_{g\downarrow}.$$
(7.9)

The H_{sd} does not change, and $H'_{sd} = H_{sd}$, while H'_m is given by

$$H'_m = -\frac{1}{2} \sum_{ff'} I_{ff'} (\mathbf{S}_f \mathbf{S}_{f'}) - \frac{1}{2} \sum_{gg'} I_{gg'} (\mathbf{S}_g \mathbf{S}_{g'})$$
$$- g\mu_B H \left\{ \sum_f S^z_f(\theta) + \sum_g S^z_g(\theta) \right\}$$
$$+ \sum_{fg} K_{fg} \left\{ \cos 2\theta \left(S^x_f S^x_g + S^z_f S^z_g \right) + S^y_f S^y_g + \sin 2\theta \left(S^z_f S^x_g - S^x_f S^z_g \right) \right\}.$$
(7.10)

Besides usual spin-conserving interatomic hopping in the non-colinear state, the unusual interatomic hopping accompanied with spin-flip process also appears.

7.3 Construction of the Atomic Representation Basis

We assume that the electron band is narrow, $|t_{fg}| \ll |A_{sd}|$. So, the s–d exchange interaction should be included in the intrasite part of the Hamiltonian which is written as

$$H_{\text{ion}} = \sum_f \{-A(\mathbf{S}_f \boldsymbol{\sigma}_f) - \bar{H} S^z_f - h\sigma^z_f - h_\perp \sigma^x_f + U n_{f\uparrow} n_{f\downarrow}\}$$
$$+ \sum_g \{-A(\mathbf{S}_g \boldsymbol{\sigma}_g) - \bar{H} S^z_g - h\sigma^z_g + h_\perp \sigma^x_g + U n_{g\uparrow} n_{g\downarrow}\}. \quad (7.11)$$

Here $\bar{H}$ is the effective self-consistent field given by

$$\bar{H} = g\mu_b H \cos\theta + I_o R - K_o R \cos 2\theta, \quad (7.12)$$

where R is the average value of the LMM

$$R = \frac{1}{N} \sum_f \langle S_f^z \rangle, \qquad (7.13)$$

and $I_0 = I(q = 0)$, $K_0 = K(q = 0)$. The longitudinal and transverse components of the magnetic field affecting holes in Eq. (7.11) are given by

$$h = 2\mu_B H \cos\theta, \quad h_\perp = -2\mu_B H \sin\theta, \qquad (7.14)$$

with the itinerant electron spin operators in the following form

$$\sigma_f^z = \sum_\sigma \sigma c_{f\sigma}^+ c_{f\sigma}, \quad \sigma_f^x = \frac{1}{2} \sum_\sigma c_{f\bar\sigma}^+ c_{f\sigma}. \qquad (7.15)$$

As described in Chapter 3, in diagonalizing the local Hamiltonian (7.11) we consider different multielectron states of the ion. Due to inequalities

$$|A| \gg \bar{H}, \quad |A| \gg |h|, \qquad (7.16)$$

the perturbation theory may be used with respect to the small parameters $\bar{H}/|A| \ll 1$, $|h|/|A| \ll 1$. The essential states are:

(1) LMM with spin S and its projection $|M\rangle$ without itinerant holes, we denote these states $|M\rangle$.
(2) sd^n configurations with one hole.

As to the two hole states, we assume that $U \to \infty$ and these states are pushed out to the region of very high energy and can be neglected. The sd^n configuration may be the high-spin $(S+1/2)$-spin states for $A_{sd} > 0$ or the low spin $S-1/2$ states for $A_{sd} < 0$. The eigenvectors of the H_{sd} for $A > 0$ can be written as (here $n = 1, 2, \ldots, 2S + 2$):

$$|n\rangle = \sqrt{\frac{2S - n + 2}{2S + 1}} |S - n + 1, \uparrow\rangle + \sqrt{\frac{n - 1}{2S + 1}} |S - n + 2, \downarrow\rangle, \qquad (7.17)$$

where $|M, \sigma\rangle$ is the sd^n configuration with an LMM state $|M\rangle$ and a hole in state $|\sigma\rangle$. It is clear that the states (7.17) are the eigenstates of H_{sd}

$$-A(\mathbf{S} \cdot \boldsymbol{\sigma})|n\rangle = -(AS/2)|n\rangle. \qquad (7.18)$$

For the case $A < 0$ the sd^n configuration eigenstates are given by

$$|n\rangle = \sqrt{\frac{2S - n + 1}{2S + 1}} |S - n + 1, \downarrow\rangle - \sqrt{\frac{n}{2S + 1}} |S - n, \uparrow\rangle, \qquad (7.19)$$

with $n = 1, 2, \ldots, 2S$. We consider below only the $A < 0$ case, the $A > 0$ case can be treated in the same way. Let us denote the following single ion Hamiltonian by H_0:

$$H_0 = -A(\boldsymbol{S} \cdot \boldsymbol{\sigma}) - \bar{H} S^z - h \sigma^z. \quad (7.20)$$

Due to the fields $\bar{H}$ and h, the spin degeneracy of the single ion eigenstates is removed. It is possible to write down the exact eigenenergies of H_0 that are given by a rather complicated expression but they can be written in a simpler and clearer form using the inequalities (7.16). To first order in the small parameters $\bar{H}/A$ and h/A, the eigenenergies of the Hamiltonian H_0 are equal to

$$\varepsilon_n = \langle n|H_0|n\rangle = \varepsilon_1 + \varepsilon(n-1), \quad (7.21)$$

$$\varepsilon_1 = \frac{A(S+1)}{2} - \bar{H}\left(S - \frac{1}{2S+1}\right) + \frac{h}{2}\frac{2S-1}{2S+1},$$

$$\varepsilon = \frac{2S+2}{2S+1}\bar{H} - \frac{h}{2S+1}. \quad (7.22)$$

In the canted phase the total single ion Hamiltonian is given by (for the F sublattice)

$$H_F = H_0 + v\sum_\sigma c_{\bar{\sigma}}^+ c_\sigma, \quad (7.23)$$

where v is equal to $\mu_B H \sin\theta$. Since the value of v is small, $(v \ll \bar{H})$, the last term in Eq. (7.23) will be treated by perturbation theory. The eigenstates $|\Psi_n\rangle$ for the operator H_F are given by

$$H_F|\Psi_n\rangle = E_n|\Psi_n\rangle. \quad (7.24)$$

To the second order in v/ε, these eigenstates may be written as

$$|\psi_n\rangle = \left\{1 - \frac{1}{2}\left(\frac{v}{\varepsilon}\right)^2 \frac{n(2S-n) + (n-1)(2S-n+1)}{(2S+1)^2}\right\}|n\rangle$$

$$+ \left(\frac{v}{\varepsilon}\right)\frac{\sqrt{n(2S-n)}}{2s+1}|n+1\rangle - \left(\frac{v}{\varepsilon}\right)\frac{\sqrt{(n-1)(2S-n+1)}}{2S+1}|n-1\rangle$$

$$+ \frac{1}{2}\left(\frac{v}{\varepsilon}\right)^2\frac{\sqrt{n(n+1)(2S-n)(2S-n+1)}}{(2S+1)^2}|n+2\rangle$$

$$+ \frac{1}{2}\left(\frac{v}{\varepsilon}\right)^2\frac{\sqrt{(n-1)(n-2)(2S-n+1)(2S-n+2)}}{(2S+1)^2}|n-2\rangle.$$

$$(7.25)$$

The second order contribution should be taken into account because near the spin-flip transition when $H \to H_c$ the ratio $H/H_c = \cos\theta$ is equal to $H/H_c \approx 1 - \theta^2/2$.

The states $|M\rangle$ and $|\Psi_n\rangle$ gives the atomic basis and are used in the definition of the Hubbard operators

$$X^{Mn} = |M\rangle\langle\Psi_n|. \tag{7.26}$$

The Fermi annihilation operator is given by

$$c_{f\sigma} = \sum_{nM} \gamma_\sigma(M,n) X_f^{Mn}, \tag{7.27}$$

with the matrix elements γ_σ calculated up to the second order in v/ε and presented in the Table 7.1.

Table 7.1 The matrix elements of the Fermi operator at $A < 0$ up to the $(v/\varepsilon)^2$ order

$\gamma_\uparrow(S-n,n)$	$-\sqrt{\dfrac{n}{2S+1}}\left\{1 - \dfrac{1}{2}\left(\dfrac{v}{\varepsilon}\right)^2 \dfrac{n(2S-n)+(n-1)(2S-n+1)}{(2S+1)^2}\right\}$
$\gamma_\uparrow(S-n-1,n)$	$-\left(\dfrac{v}{\varepsilon}\right)\dfrac{\sqrt{n(2S-n)(n+1)}}{(2S+1)^{3/2}}$
$\gamma_\uparrow(S-n+1,n)$	$\left(\dfrac{v}{\varepsilon}\right)\dfrac{(n-1)\sqrt{2S-n+1}}{(2S+1)^{3/2}}$
$\gamma_\uparrow(S-n-2,n)$	$-\dfrac{1}{2}\left(\dfrac{v}{\varepsilon}\right)^2 \dfrac{\sqrt{n(n+1)(n+2)(2S-n)(2S-n-1)}}{(2S+1)^{3/2}}$
$\gamma_\uparrow(S-n+2,n)$	$-\dfrac{1}{2}\left(\dfrac{v}{\varepsilon}\right)^2 \dfrac{(n-2)\sqrt{(n-1)(2S-n+1)(2S-n+2)}}{(2S+1)^{3/2}}$
$\gamma_\downarrow(S-n+1,n)$	$-\sqrt{\dfrac{2S-n+1}{2S+1}}\left\{1 - \dfrac{1}{2}\left(\dfrac{v}{\varepsilon}\right)^2 \dfrac{n(2S-n)+(n-1)(2S-n+1)}{(2S+1)^2}\right\}$
$\gamma_\downarrow(S-n,n)$	$\left(\dfrac{v}{\varepsilon}\right)\dfrac{(2S-n)\sqrt{n}}{(2S+1)^{3/2}}$
$\gamma_\downarrow(S-n+2,n)$	$-\left(\dfrac{v}{\varepsilon}\right)\dfrac{\sqrt{(n-1)(2S-n+1)(2S-n+2)}}{(2S+1)^{3/2}}$
$\gamma_\downarrow(S-n-1,n)$	$\dfrac{1}{2}\left(\dfrac{v}{\varepsilon}\right)^2 \dfrac{(2S-n-1)\sqrt{n(n+1)(2S-n)}}{(2S+1)^{3/2}}$
$\gamma_\downarrow(S-n-3,n)$	$\dfrac{1}{2}\left(\dfrac{v}{\varepsilon}\right)^2 \dfrac{\sqrt{(n-1)(n-2)(2S-n+1)(2S-n+2)(2S-n+3)}}{(2S+1)^{3/2}}$

7.4 The Hamiltonian of the Narrow Band Antiferromagnet in the Non-colinear Phase in Atomic Representation

The two sublattice Hamiltonian describes a lattice of ions in the d^n and sd^n configurations

$$H' = H'_{\text{ion}} + H_{\text{int}}, \qquad (7.28)$$

where the local part is given by

$$H'_{\text{ion}} = \sum_{fM} E_M X_f^{MM} + \sum_{gM} E_M X_g^{MM} + \sum_{fn}(E_n - \mu) X_f^{nn} + \sum_{gn}(E_n - \mu) X_g^{nn}. \qquad (7.29)$$

The energy of the sd^n terms is equal to

$$E_n = \varepsilon_n + \frac{2n - 2S - 1}{(2S+1)^2}\left(\frac{v^2}{\varepsilon}\right), \quad n = 1, 2, \ldots, 2S. \qquad (7.30)$$

The interaction Hamiltonian is written as

$$H_{\text{int}} = \sum_{\substack{AA' \\ l' \in A'}} \sum_{l \in A} \sum_{\alpha\beta\sigma} V_\sigma^{\alpha\beta}(l, l') X_l^{-\alpha} X_{l'}^{\beta}. \qquad (7.31)$$

The single ion eigenproblem for the G-sublattice can be obtained from the F-sublattice by substitution $\theta \to -\theta$. The electron matrix elements in Eq. (7.27) should be written as $\gamma_\sigma(\theta, \alpha)$ for the F-sublattice and $\gamma_\sigma(-\theta, \alpha)$ for the G-sublattice. For simplicity we denote $\gamma_\sigma(\theta, \alpha) = \gamma_\sigma(\alpha)$ and $\gamma_\sigma(-\theta, \alpha) = \bar{\gamma}_\sigma(\alpha)$. Then, the X-representation for the Fermi operators is the following

$$c_{f\sigma} = \sum_\alpha \gamma_\sigma(\alpha) X_f^\alpha, \qquad c_{g\sigma} = \sum_\alpha \bar{\gamma}_\sigma(\alpha) X_g^\alpha. \qquad (7.32)$$

The matrix elements of the interaction are equal to

$$\begin{aligned}
V_\sigma^{\alpha\beta}(f, f') &= t_{f,f'} \gamma_\sigma(\alpha)\gamma_\sigma(\beta), \\
V_\sigma^{\alpha\beta}(g, g') &= t_{g,g'} \bar{\gamma}_\sigma(\alpha)\bar{\gamma}_\sigma(\beta), \\
V_\sigma^{\alpha\beta}(f, g) &= t_{f,g}[\cos\theta \gamma_\sigma(\alpha)\bar{\gamma}_\sigma(\beta) + 2\sigma \sin\theta \gamma_\sigma(\alpha)\bar{\gamma}_{\bar\sigma}(\beta)], \\
V_\sigma^{\alpha\beta}(g, f) &= t_{f,g}[\cos\theta \bar{\gamma}_\sigma(\alpha)\gamma_\sigma(\beta) + 2\sigma \sin\theta \bar{\gamma}_\sigma(\alpha)\gamma_{\bar\sigma}(\beta)].
\end{aligned} \qquad (7.33)$$

7.5 The Green's Functions and Dispersion Relations

We introduce the following Green's function

$$G_{\alpha\beta}^{AA'}(l\tau,l'\tau') = -\left\langle T_\tau \tilde{X}_l^\alpha(\tau)\tilde{X}_{l'}^{-\beta}(\tau')\right\rangle, \qquad (7.34)$$

where the site $l(l')$ belongs to the sublattice $A(A')$. From the Fourier transform of the Green's function (7.34) in the mean field approximation, the following diagram equation may be obtained

$$\begin{array}{c} A \quad A' \\ \xrightarrow{} \\ \alpha \quad \beta \end{array} = \begin{array}{c} A \quad A' \\ \xrightarrow{}\!\!\circ \\ \alpha \quad \beta \end{array} + \begin{array}{c} A \quad\quad A_1 \quad A' \\ \xrightarrow{}\!\!\alpha\!\!\sim\!\!\sim\!\!\xrightarrow{} \\ \alpha \qquad \alpha_1 \quad \beta \end{array}. \qquad (7.35)$$

Here the bare line is given by the intra-ionic propagator $\begin{array}{c} A \\ \xrightarrow{} \\ \alpha \end{array}$

$$G_\alpha(\omega_n) = [i\omega_n + \alpha E]^{-1}, \quad \omega_n = (2n+1)\pi T, \qquad (7.36)$$

with the scalar product $\boldsymbol{\alpha}\cdot\boldsymbol{E}$ being equal to

$$\boldsymbol{\alpha}\cdot\boldsymbol{E} \equiv \boldsymbol{\alpha}(m,n)\cdot\boldsymbol{E} = E_m - E_n.$$

The interaction shown by the wavy line in Eq. (7.35) is the matrix with sublattice indexes, for example the inter-sublattice hopping is described by the expression

$$\Gamma_{\boldsymbol{q}}^{\alpha\alpha_1} = \sum_\sigma \Gamma_{\boldsymbol{q}}[\cos\theta\gamma_\sigma(\alpha)\gamma_{\bar\sigma}(\alpha_1) + 2\sigma\sin\theta\gamma_\sigma(\alpha)\bar\gamma_{\bar\sigma}(\alpha_1)], \qquad (7.37)$$

with

$$\Gamma_{\boldsymbol{q}} = \frac{1}{N}\sum_g t_{fg}\exp\{i\boldsymbol{q}\cdot(\boldsymbol{R}_f - \boldsymbol{R}_g)\}. \qquad (7.38)$$

The intra-sublattice hopping is given by

$$t_{\boldsymbol{q}}^{\alpha\alpha_1} = \sum_\sigma t_{\boldsymbol{q}}\gamma_\sigma(\alpha)\gamma_\sigma(\alpha_1) \qquad (7.39)$$

with

$$t_{\boldsymbol{q}} = \frac{1}{N}\sum_{f'} t_{ff'}\exp\{i\boldsymbol{q}\cdot(\boldsymbol{R}_f - \boldsymbol{R}_{f'})\}. \qquad (7.40)$$

The analytical form of Eq. (7.35) is the following

$$G^{FF}_{\alpha\beta}(k,\omega_n) = \delta_{\alpha\beta}G_\alpha(\omega_n)b(\alpha) + G_\alpha(\omega_n)b(\alpha)$$
$$\times \sum_{\alpha_1} \left\{ t^{\alpha\alpha_1}_k G^{FF}_{\alpha_1\beta}(k,\omega_n) + \Gamma^{\alpha\alpha_1}_k G^{GF}_{\alpha_1\beta}(k,\omega_n) \right\}, \quad (7.41)$$

$$G^{GF}_{\alpha\beta}(k,\omega_n) = G_\alpha(\omega_n)b(\alpha) \left\{ t_k \sum_{\sigma\alpha_1} \bar{\gamma}_\sigma(\alpha)\bar{\gamma}_\sigma(\alpha_1) G^{GF}_{\alpha_1\beta}(k,\omega_n) \right.$$

$$+ \left[\Gamma_k \cos\theta \sum_{\sigma\alpha_1} \bar{\gamma}_\sigma(\alpha)\gamma_\sigma(\alpha_1) \right.$$

$$\left. + \Gamma_k \sin\theta \sum_{\sigma\alpha_1} 2\sigma\bar{\gamma}_\sigma(\alpha)\gamma_\sigma(\alpha_1) \right] G^{FF}_{\alpha_1\beta}(k,\omega_n). \quad (7.42)$$

The inter-ionic interaction matrix elements $V^{\alpha\beta}$ in Eq. (7.33) have a multiplicative structure, being a product or a sum of the products of separate functions of only α and β. This structure allows one to simplify the solution of Eqs. (7.41, 7.42) for an arbitrary value of the spin S. We introduce the following functions

$$Y^{FF}_{\sigma\beta} = \sum_{\alpha_1} \gamma_\sigma(\alpha_1) G^{FF}_{\alpha_1\beta}, \qquad Y^{GF}_{\sigma\beta} = \sum_{\alpha_1} \bar{\gamma}_\sigma(\alpha_1) G^{GF}_{\alpha_1\beta}. \quad (7.43)$$

The explicit expression for Y^{FF} comes from Eq. (7.41) and can be written in the form

$$Y^{FF}_{\sigma\beta} = \gamma_\sigma(\beta)G_\beta(\omega_n) + t_k \sum_{\sigma_1} L^F_{\sigma\sigma_1}(\omega_n) Y^{FF}_{\sigma_1\beta} + \Gamma_k \sum_{\sigma_1} M^F_{\sigma\sigma_1} Y^{GF}_{\sigma_1\beta}, \quad (7.44)$$

with the following notations

$$L^F_{\sigma\sigma_1}(\omega_n) = \sum_\alpha \gamma_\sigma(\alpha)\gamma_{\sigma_1}(\alpha) G_\alpha(\omega_n) b(\alpha),$$
$$M^F_{\sigma\sigma_1}(\omega_n) = L^F_{\sigma\sigma_1}(\omega_n) \cos\theta - (2\sigma_1) L^F_{\sigma\bar{\sigma}_1}(\omega_n) \sin\theta. \quad (7.45)$$

A similar equation for the Y^{GF} is given by

$$Y^{GF}_{\sigma\beta} = t_k \sum_{\sigma_1} L^G_{\sigma\sigma_1}(\omega_n) Y^{GF}_{\sigma_1\beta} + \Gamma_k \sum_{\sigma_1} M^G_{\sigma\sigma_1} Y^{FF}_{\sigma_1\beta} \quad (7.46)$$

with notations

$$L^G_{\sigma\sigma_1}(\omega_n) = \sum_\alpha \bar{\gamma}_\sigma(\alpha)\bar{\gamma}_{\sigma_1}(\alpha)G_\alpha(\omega_n)b(\alpha),$$
$$M^G_{\sigma\sigma_1}(\omega_n) = L^G_{\sigma\sigma_1}(\omega_n)\cos\theta + (2\sigma_1)L^G_{\sigma\bar\sigma_1}(\omega_n)\sin\theta. \quad (7.47)$$

Thus, instead of a large number of equations (7.41–7.42) determined by the spin value S we have obtained a system of four equations for the functions $Y^{FF}_{\uparrow\beta}$, $Y^{FF}_{\downarrow\beta}$, $Y^{GF}_{\uparrow\beta}$ and $Y^{GF}_{\downarrow\beta}$. The initial Green's functions G^{FF}, G^{GF} can be easily expressed through the Y-functions.

The dispersion equation is determined by zeros of the Y-function denominator which is given by the determinant

$$\Delta(\boldsymbol{k},i\omega_n) = \begin{vmatrix} 1 - t_{\boldsymbol{k}}L^F_{\uparrow\uparrow} & -\Gamma_{\boldsymbol{k}}M^F_{\uparrow\uparrow} & -t_{\boldsymbol{k}}L^F_{\uparrow\downarrow} & -\Gamma_{\boldsymbol{k}}M^F_{\uparrow\downarrow} \\ -\Gamma_{\boldsymbol{k}}M^G_{\uparrow\uparrow} & 1 - t_{\boldsymbol{k}}L^G_{\uparrow\uparrow} & -\Gamma_{\boldsymbol{k}}M^G_{\uparrow\downarrow} & -t_{\boldsymbol{k}}L^G_{\uparrow\downarrow} \\ -t_{\boldsymbol{k}}L^F_{\downarrow\uparrow} & -\Gamma_{\boldsymbol{k}}M^F_{\downarrow\uparrow} & 1 - t_{\boldsymbol{k}}L^F_{\downarrow\downarrow} & -\Gamma_{\boldsymbol{k}}M^F_{\downarrow\downarrow} \\ -\Gamma_{\boldsymbol{k}}M^G_{\downarrow\uparrow} & -t_{\boldsymbol{k}}L^G_{\downarrow\uparrow} & -\Gamma_{\boldsymbol{k}'}M^G_{\downarrow\downarrow} & 1 - t_{\boldsymbol{k}'}L^G_{\downarrow\downarrow} \end{vmatrix}. \quad (7.48)$$

Due to the sublattice symmetry

$$L^G_{\sigma\sigma} = L^F_{\sigma\sigma}, \quad L^G_{\sigma\bar\sigma} = -L^F_{\sigma\bar\sigma}, \quad M^G_{\sigma\sigma} = M^F_{\sigma\sigma}, \quad M^G_{\sigma\bar\sigma} = -M^F_{\sigma\bar\sigma}, \quad (7.49)$$

we use only the F-sublattice notations and omit the sublattice index below. Then, the determinant (7.48) is given by

$$\begin{aligned}\Delta(\boldsymbol{k},\omega_n) =& \left[(1 - t_{\boldsymbol{k}}L_{\uparrow\uparrow})^2 - \Gamma^2_{\boldsymbol{k}}M^2_{\uparrow\uparrow}\right]\left[(1 - t_{\boldsymbol{k}}L_{\downarrow\downarrow})^2 - \Gamma^2_{\boldsymbol{k}}M^2_{\downarrow\downarrow}\right] + \Gamma^4_{\boldsymbol{k}}M^2_{\uparrow\downarrow}M^2_{\downarrow\uparrow} \\ &+ 2(\Gamma^2_{\boldsymbol{k}}M_{\uparrow\downarrow}M_{\downarrow\uparrow} - t^2_{\boldsymbol{k}}L^2_{\uparrow\downarrow})\left[(1 - t_{\boldsymbol{k}}L_{\uparrow\uparrow})(1 - t_{\boldsymbol{k}}L_{\downarrow\downarrow}) - \Gamma^2_{\boldsymbol{k}}M_{\uparrow\uparrow}M_{\downarrow\downarrow}\right] \\ &+ 2t_{\boldsymbol{k}}\Gamma^2_{\boldsymbol{k}}L_{\uparrow\downarrow}(M_{\uparrow\downarrow} - M_{\downarrow\uparrow})(M_{\uparrow\uparrow} - M_{\downarrow\downarrow}) \\ &+ t^2_{\boldsymbol{k}}L^2_{\uparrow\downarrow}\left[t^2_{\boldsymbol{k}}L^2_{\uparrow\downarrow} - \Gamma^2_{\boldsymbol{k}}\left(M^2_{\uparrow\downarrow} + M^2_{\downarrow\uparrow}\right)\right]. \end{aligned} \quad (7.50)$$

The solutions of the dispersion equation

$$\Delta(\boldsymbol{k}, i\omega_n \to \omega + i\delta) = 0$$

give the quasiparticle band structure of the narrow band antiferromagnet which depends on the occupation numbers of d^n configuration states

$$N_M = \langle X^{M,M}\rangle \quad (M = S, S-1, \ldots, -S+1, -S)$$

and sd^n configuration states

$$N_n = \langle X^{n,n}\rangle \quad (n = 1, 2, \ldots, 2S).$$

At high temperatures $T \sim T_N$ and rather high electron concentration the number of essential states making contribution to Eq. (7.50) is not small and the band structure is quite complicated.

For the dHvA effect we will consider only the low temperature limit $T \ll T_N$ when only the ground states of the d^n ion with occupation N_S and sd^n ion with occupation N_1 need to be taken into account.

7.6 The Spectrum of Holes Near the Spin-flip Transition

Using the matrix elements of the Fermi operators from the Table 7.1 we can write the elements of Eq. (7.50) in the form

$$L_{\uparrow\uparrow}(\omega_n) = \left(\frac{N_1}{2S+1}\right)\Lambda(S-1,1) + \xi\left(\frac{v}{\varepsilon}\right) N_1[2\Lambda(S-2,1) - \Lambda(S-1,1)],$$

$$L_{\downarrow\downarrow}(\omega_n) = \left(\frac{2S}{2S+1}\right)\Lambda(S,1) + \xi\left(\frac{v}{\varepsilon}\right)$$
$$\times [2SN_S\Lambda(S,2) + (2S-1)N_1\Lambda(S-1,1) - 2SN_1\Lambda(S,1)],$$

$$L_{\uparrow\downarrow}(\omega_n) = L_{\downarrow\uparrow}(\omega_n) = -N_1\xi_1\Lambda(S-1,1),$$

$$M_{\uparrow\uparrow}(\omega_n) = L_{\uparrow\uparrow}(\omega_n)\cos\theta + N_1\xi\sin\theta\Lambda(S-1,1),$$

$$M_{\downarrow\downarrow}(\omega_n) = L_{\downarrow\downarrow}(\omega_n)\cos\theta - N_1\xi\sin\theta\Lambda(S-1,1),$$

$$M_{\uparrow\downarrow}(\omega_n) = \left(\frac{N_1\sin\theta}{2S+1}\right)\Lambda(S-1,1) - N_1\xi\cos\theta\Lambda(S-1,1),$$

$$M_{\downarrow\uparrow}(\omega_n) = -\left(\frac{2S}{2S+1}\sin\theta\right)\Lambda(S,1) - N_1\xi\cos\theta\Lambda(S-1,1),$$

(7.51)

where

$$\xi = \left(\frac{v}{\varepsilon}\right)\frac{2S-1}{(2S+1)^2}, \qquad \Lambda(M,l) = \frac{1}{i\omega_n + E_M - E_l}.$$

The following estimations are clear from Eq. (7.51):

$$L_{\sigma\sigma} \sim (v/\varepsilon)^0, \qquad M_{\sigma\sigma} \sim (v/\varepsilon)^0,$$
$$L_{\sigma\bar{\sigma}} \sim (v/\varepsilon)^1, \qquad M_{\sigma\bar{\sigma}} \sim (v/\varepsilon)^1,$$

(7.52)

and the dispersion relation up to the second order in θ may be written in the following form

$$\left(1 - \Gamma_k^2 M_{\uparrow\uparrow}^2\right)\left(1 - \Gamma_k^2 M_{\downarrow\downarrow}^2\right) + 2\Gamma_k^2 M_{\uparrow\downarrow}M_{\downarrow\uparrow}\left(1 - \Gamma_k^2 M_{\uparrow\uparrow}M_{\downarrow\downarrow}\right) = 0. \quad (7.53)$$

Here we consider only the nearest neighbor hopping. The quasiparticle dispersion law is given by

$$E(k) = \frac{A(S+1)}{2} + \frac{\bar{H}}{2S+1} + \frac{h}{2}\frac{2S-1}{2S+1} - \frac{2S}{2S+1}|\Gamma_k|\cos\theta$$

$$+ N_1 \left\{ \frac{2S-1}{(2S+1)^3}\left(\frac{v}{\varepsilon}\right)^2 + \frac{2(2S-1)}{(2S+1)^2}\left(\frac{v\sin\theta}{\varepsilon}\right) \right.$$

$$\left. - \frac{2S-1}{(2S+1)^2}\left(\frac{v^2}{\varepsilon}\right) - \frac{\sin^2\theta}{2S+1} \right\} |\Gamma_k|. \tag{7.54}$$

7.7 The Peculiarities of the de Haas-van Alphen Effect in Antiferromagnetic Semimetal with Magnetic Polaron States

According to the discussion of the electronic structure of *CeAs*, *CeBi* and *CeP* in the beginning of this chapter, we model these semimetals with the conductivity band in the effective mass approximation

$$E_{e\sigma}(k) = \frac{\hbar^2 k^2}{2m_e} - 2\mu_B H\sigma, \qquad \sigma = \pm 1/2, \tag{7.55}$$

while the magnetopolaron states at the top of the valence band are given by Eq. (7.54). The hole energy near the extreme point in the Brillouin zone may be written as

$$E_h(k) = -\Delta(H) + \frac{\hbar^2 k^2}{2m_{hc}}\cos\theta, \tag{7.56}$$

where m_{hc} is the effective mass of the magnetopolaron hole at the point of spin-flip transition. If the free hole mass is m_h, then the magnetopolaron mass is equal to $m_{hc} = m_h(2S+1)/2S$ (Nagaev 1983). It should be emphasized that the dispersion of holes given by Eq. (7.56) is valid only at $H \sim H_c$. If $H \ll H_c$ and $\cos\theta \ll 1$, the excited sd^n ion states are important and the magnetopolaron bandwidth is given by $t^2/|A|$ instead of $|t|\cos\theta$ (Nagaev 1983). Near the spin-flip transition the bandwidth is proportional to t and the t^2/A contributions may be neglected.

The value $\Delta(H)$ in Eq. (7.56) is equal to

$$\Delta(H) = \Delta_c + W_s(\cos\theta - 1) + \mu_B(H_c - H\cos\theta) - \frac{2K_oS}{2S+1}\sin^2\theta \tag{7.57}$$

and at the spin-flip transition $\Delta = \Delta_c$. The bottom of the magnetopolaron band reveals a large shift when the magnetic field is decreased. The Zeeman shift is $\mu_B H_c$, while the shift due to the change of the canting angle in the magnetic field is $W_S \gg \mu_B H_c$. Here, $W_S = 16S \left|t_1^h\right|/(2S+1)$ is the spin polaron bandwidth at the spin-flip point when $H = H_c$. This difference is very important for the dHvA effect. In the colinear phase at $H > H_c$ the expression (7.57) is valid, provided $\cos\theta = 1$.

In the case of the low concentration of carriers it is possible to neglect its influence on the orientation of LMM. In this case (Borovik-Romanov 1962; Gurevich 1973), we have

$$\cos\theta = g\mu_B H/2SK_o. \tag{7.58}$$

Due to the Landau quantization in strong magnetic field (Landau and Lifshits 1974; Lifshits et al 1971) the oscillating part of the itinerant electron magnetization may be written in the following form

$$M_{\sim}^{el} = \frac{T(\hbar m_e \omega_e)^{3/2}}{2\pi \hbar^3} \left[\frac{d}{dH}\left(\frac{\mu}{\hbar \omega_e}\right)\right]$$
$$\times \sum_{k=1,\sigma}^{\infty} \frac{(-1)^k}{\sqrt{k}} \frac{\sin(\varphi(H) + 2\pi k \sigma m_e/m_0 - \pi/4))}{\sin h(2\pi^2 kT/\hbar \omega_n)}, \tag{7.59}$$

where $\varphi(H) = 2\pi k\mu/\hbar\omega_e$, $\omega_e = eH/mc$ is a cyclotron frequency and m_0 is a free electron mass. The electron neutrality condition results in the following dependence of the chemical potential on the magnetic field

$$\mu(H) = \frac{\Delta(H)}{1+\chi\cos\theta}, \quad \chi = 2^{2/3} m_e/m_{hc}. \tag{7.60}$$

To study the peculiarities of the dHvA effect we consider the derivatives of the phase with respect to the magnetic field above the spin-flip transition

$$\left[\frac{d}{dH}\varphi^+(H)\right]_{H=H_c} = \frac{2\pi k \mu_c}{\hbar \omega_e H_c}, \quad \mu_c = \frac{\Delta_c}{1+\chi},$$

and below the transition

$$\left[\frac{d}{dH}\varphi^-(H)\right]_{H=H_c} = -2\pi k \frac{\mu_c - H_c(\partial\mu/\partial H)_c}{\hbar \omega_e H_c}.$$

Using Eqs. (7.57) and (7.60), one can obtain

$$\left[\frac{d}{dH}\varphi^-(H)\right]_{H=H_c} = -2\pi k \left\{\frac{\mu_c(1+2\chi)-W_s}{1+\chi}\right\} \cdot \left(\frac{1}{\hbar\omega_e H_c}\right).$$

The comparison of phases below and above the transition shows that φ^+ decreases and φ^- increases with the increasing of H. The increase of φ^- is clear from the inequality

$$\mu_c(1+2\chi) - W_s < 0,$$

that is valid for small value of μ_c at $n \ll 1$. The phases below and above the transition are equal and the dHvA period does not change at some critical value of the chemical potential

$$\mu_{cr} = W_s/(2+3\chi).$$

The corresponding values for the spin polaron band shift Δ_{cr} and the carrier concentration n_{cr} are given by

$$\Delta_{cr} = \frac{1+\chi}{2+3\chi} W_s, \qquad n_{cr} = \frac{8\sqrt{2}}{3\pi^2}\left(\frac{\chi}{2+3\chi}\right)^{3/2}.$$

At the concentration $n < n_{cr}$ with chemical potential $\mu < \mu_{cr}$ there is a sharp change of the period. At $H < H_c$ the frequency of oscillations are higher at $H > H_c$ (Fig. 7.1). The dHvA oscillations were obtained here with the following values of the parameters $T_N = 7\,\mathrm{K}$, $T = 1\,\mathrm{K}$, $S = 5/2$, $g = 2$, $m_e = m_0$, $m_h = 10\,m_e$. The carrier concentration is equal to $n = 2.3 \times 10^{-3}$, while $n_{cr} = 6.2 \times 10^{-3}$. The dHvA oscillations for the case of higher carrier concentration ($n = 1.02 \times 10^{-2}$) is shown in the Fig. 7.2 keeping fixed the values of the other parameters. It is clear from Figs. 7.1 and 7.2 that the amplitude of oscillations changes due to the factor $d(\mu/\hbar\omega_e)/dH$ in

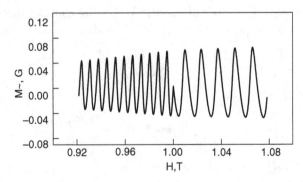

Fig. 7.1 The change of the oscillation period of the antiferromagnet semimetal with carrier concentration $n < n_{cr}$.

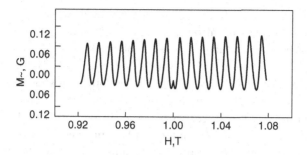

Fig. 7.2 The change of the oscillation period for $n > n_{cr}$.

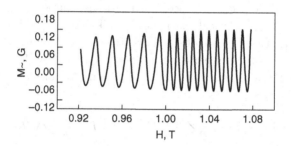

Fig. 7.3 The dHvA oscillations at $n = n_{cr}$.

Eq. (7.59). At $n = n_{cr}$ there is a phase shift but the period of the dHvA oscillations does not change (Fig. 7.3).

The hole contribution $M_\sim^h$ to the oscillation part of the magnetization is small due to large hole mass $m_h \gg m_e$, so the electron contribution (7.59) is the main one. Nevertheless it should be mentioned that there is a strong dependence of the hole spin polaron band on the magnetic field that results in the non-Fermi liquid anomaly of the dHvA effect near the spin-flip transition.

7.8 The Temperature Quantum Oscillations Caused by Non-Fermi Liquid Effects

Another non-Fermi liquid effect in magnetic semiconductor which results in the temperature quantum oscillations (TQO) occurs in the doped semiconductor with the Fermi level close to the local excitation energy

$$\Omega_{pq} = E_{n+1}(q) - E_n(p),$$

where $E_n(p)$ is the n-electron term $|p\rangle$. For example, in the n-type chromium spinel HgCr$_2$Se$_4$ the relevant terms are $d^3(Cr^{+3})$ and $d^4(Cr^{2+})$. The proper model for such semiconductor is the combination of the periodic Anderson model and $s-d$ exchange model (Gavrichkov et al 1986; Ovchinnikov 1991). Due to the $s-d$ exchange interaction J, the opposite spin subbands split in the ferromagnetic phase while the $s-d$ hybridization results in a mixing of the itinerant electron states with the localized d-type excitations Ω. In a degenerate magnetic semiconductor with a single electron band structure (without the Ω level), the temperature induced shift of the band energy is accompanied by a similar shift of the chemical potential with negligible Fermi liquid renormalization $\delta\mu \sim T^2/\varepsilon_F$. But in the case of $\varepsilon_F \sim \Omega$ there is a typical mixed valence process of the redistribution of itinerant and localized electrons. The calculations of Val'kov and Dzebisashvili (1997) have shown that in the case of small effective mass of the itinerant electron $m^* \sim 10^{-2}m_e$ in the temperature interval $\mu_B H \ll T \ll \hbar\omega_c$, the relative chemical potential measured from the bottom of the conductivity band can be written as

$$\mu(T, H) = \mu(0) - \frac{J}{2}Z(3/2)t^{3/2} + Jt\sqrt{h} - \frac{35}{96\pi}J\sqrt{th}, \qquad (7.61)$$

where $Z(3/2) = 2.612$, t and h are the dimensionless temperature and magnetic field, $t = T/4\pi IS$, $h = \mu_B H/IS$, with I being the parameter of exchange interaction between the neighboring spins and S the spin of the localized multielectron term ($S = 3/2$ for Cr^{3+}). Since in the dHvA experiment the oscillating part of the thermodynamic potential and its derivatives are determined by the factor

$$\sin\left(2\pi k \frac{\mu(T, H)}{\hbar\omega_c}\right)$$
$$= \sin\left\{\frac{2\pi k}{\hbar\omega_c}\left[\mu(0) - \frac{J}{2}Z(3/2)t^{3/2} + Jt\sqrt{h} - \frac{35}{96\pi}J\sqrt{th}\right]\right\}, \qquad (7.62)$$

it is clear that several non-Fermi liquid effects should occur. The first is the violation of the $1/H$ periodicity in the quantum oscillation at fixed temperature. However, the most unusual is the temperature dependence at the fixed magnetic field in the form of a nonperiodic temperature quantum oscillations (TQO). For comparison, the temperature dependence of the dHvA signal in the Fermi liquid case is given by the well-known Lifshits–Kosevich damping. Experimental measurements of the dHvA signal in the single crystals of HrCr$_2$Se$_4$ n-type with the record value of the electron

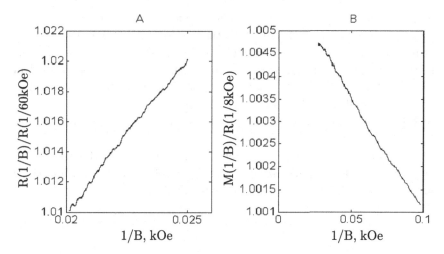

Fig. 7.4 The field dependence of the resistance (a) and the magnetization (b) in n-HgCr$_2$Se$_4$ at $T = 4.2$ K (Balaev et al 1998). 1 T = 10 kO$_e$.

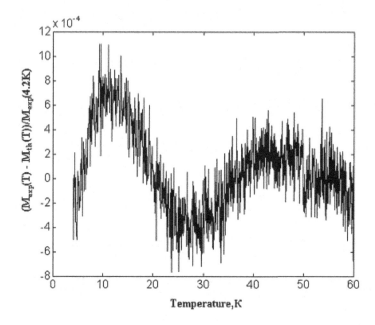

Fig. 7.5 The temperature quantum oscillations of the magnetization at $H = 6$ T in n-HgCr$_2$Se$_4$ is given by the subtraction of the calculated monotonic spin-wave contribution $M_{\text{theor}}(T)$ from the measured $M_{\exp}(T)$ (Balaev et al 1998).

mobility among spinel magnetic semiconductors $\mu \sim 10^3 \, \mathrm{cm}^2/\mathrm{V} \cdot \mathrm{s}$ at $T = 77 \, \mathrm{K}$ have revealed the TQO for the magnetic field 6T in the temperature interval 4–60 K (Ovchinnikov et al 1995). The quantum oscillations of the resistivity and magnetization in n-HgCr$_2$Se$_4$ have been studied in detail by Balaev et al (1998). Both non-Fermi liquid effects discussed above have been confirmed. The nonperiodicity in $1/H$ of R and M are shown in Fig. 7.4 and the nonperiodic TQO of magnetization are presented in Fig. 7.5.

In the antiferromagnetic semimetal described above in the Section 7.7 the same consideration also results in the TQO of the magnetization which are periodic with T^2 (Val'kov and Dzebisashvili 1998).

Chapter 8

The Electronic Structure of Copper Oxides in the Multiorbital p–d Model

The effects of SEC are known to be crucial in the formation of the electronic structure of the undoped and weakly doped cuprates. The conventional single electron approach does not treat the SEC effects adequately. The 3-band p–d-model (Emery 1987; Varma et al 1987) is the simplest model of cuprates incorporating both the SEC and the nature of the chemical bonding. In this model there is one hole per unit cell given by the CuO_4 (CuO_6) cluster of the undoped CuO_2 layer and this hole is mostly localized in Cu^{2+} ion with spin $S = 1/2$. In the hole doped systems with the hole concentration $n_h = 1 + x$ the carriers (extra holes) occupy mainly the oxygen p-orbitals. The two-hole pair "a hole on Cu^{2+} and a hole on oxygen" forms the spin singlet state (Zhang and Rice 1988) with the corresponding spin triplet state having significantly higher energy $\varepsilon_t - \varepsilon_S > 2\,\text{eV}$. This is insignificant for the low energy physics. Therefore the effective low energy model is given by reducing the three-band p–d model to the single band Hubbard model and then to the t–J model (see reviews by Dagotto 1994; Kampf 1994; Brenig 1995; Ovchinnikov 1997).

Nevertheless, there are experimental and theoretical evidence emphasizing the importance of the other d-orbitals beyond the three-band p–d model.

The polarized X-ray absorption spectroscopy (Bianconi et al 1988; Pompa et al 1991) and the electron energy loss spectroscopy (Romberg 1990) reveal quite a significant contribution of the $d_{3z^2-r^2}$ copper electrons. The theoretical consideration of the multiorbital models which incorporate $d_{3z^2-r^2}$ state for the CuO_6 cluster has shown that the two-hole triplet $^3B_{1g}$ formed with the participation of the $d_{3z^2-r^2}$ orbital lies only 0.7 eV above the singlet $^1A_{1g}$ term and the crossover between $^1B_{1g}$ and $^1A_{1g}$ terms is possible with the variation of the model parameters (Eskes et al 1990; Eskes and Sawatzky 1991). A similar conclusion has been obtained by the configuration interaction calculations for the CuO_6 cluster by Kamimura and Eto (1990) who have shown that the small reduction of the Cu–axial oxygen distance results in the crossover for $^3B_{1g}$ and $^1A_{1g}$ two hole terms. The low singlet–triplet excitation energy may result in their mixing outside the Γ point (Emery and Reiter 1998).

In this chapter we use the generalized tight-binding approach to the quasiparticle band structure in the framework of the multiband p–d model which can account for two orbitals of copper ($d_{x^2-y^2} \equiv d_x$ and $d_{3z^2-r^2} \equiv d_z$) and 2p-oxygen orbitals (Gaididei and Loktev 1988). The model of CuO_2 layer is assumed to contain the CuO_4 or CuO_6 clusters depending on the type of compound. In the case of CuO_6 cluster the apical oxygen p_z-orbitals are also included.

8.1 The Exact Diagonalization of the CuO_4 Cluster

The wavefunctions for the CuO_6 (as well as for the CuO_4) cluster are shown in Fig. 8.1. While the hopping matrix elements depend on the phase choice the result, of course, does not depend on it. The exact diagonalization of the Hamiltonian (2.41) for the CuO_4 cluster is to be done separately for each sector of the Hilbert space with the given number of holes $n_h = 0, 1, 2, \ldots$.

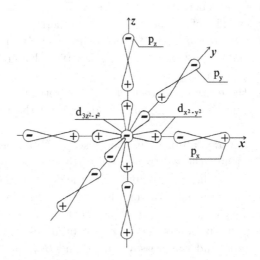

Fig. 8.1 The atomic orbitals in the CuO_6 cluster.

The Electronic Structure of Copper Oxides in the Multiorbital p–d Model   197

In the $n_h = 0$ sector there is only one singlet vacuum state $|0\rangle$ with the energy $E = 0$ corresponding to the filled shell $d^{10}p^6$. In the $n_h = 1$ sector all atomic and molecular orbitals have spin $S = 1/2$. The initial atomic basis contains the following orbitals

$$p_x(\mathbf{r} - \mathbf{a}_x/2) \equiv p_{x-1/2}, \quad p_x(\mathbf{r} + \mathbf{a}_x/2) \equiv p_{x+1/2},$$
$$p_y(\mathbf{r} - \mathbf{a}_y/2) \equiv p_{y-1/2}, \quad p_y(\mathbf{r} + \mathbf{a}_y/2) \equiv p_{y+1/2}. \quad (8.1)$$

We introduce two oxygen molecular orbitals of b_1 and a_1 symmetry:

$$|b_1\rangle = \frac{1}{2}(|p_{x-1/2}\rangle - |p_{x+1/2}\rangle + |p_{y-1/2}\rangle - |p_{y+1/2}\rangle), \quad (8.2)$$

$$|a_1\rangle = \frac{1}{2}(-|p_{x-1/2}\rangle + |p_{x+1/2}\rangle + |p_{y-1/2}\rangle - |p_{y+1/2}\rangle), \quad (8.3)$$

as well as the nonbonding orbitals

$$|n_1\rangle = \frac{1}{2}(|p_{x-1/2}\rangle + |p_{x+1/2}\rangle + |p_{y-1/2}\rangle + |p_{y+1/2}\rangle),$$
$$|n_2\rangle = \frac{1}{2}(|p_{x-1/2}\rangle + |p_{x+1/2}\rangle - |p_{y-1/2}\rangle - |p_{y+1/2}\rangle).$$

In this basis the single hole part of the Hamiltonian (2.41) is written in the form

$$H_{pd}^{(0)} = \varepsilon_{dx} d_x^+ d_x + \varepsilon_{dz} d_z^+ d_z + \sum_{i=1}^{4} \varepsilon_p p_i^+ p_i + t_{pd}(d_x^+ b_1 + \text{h.c.})$$
$$+ \frac{t_{pd}}{\sqrt{3}}(d_z^+ a_1 + \text{h.c.}) - 2t_{pp} b_1^+ b_1 + 2t_{pp} a_1^+ a_1. \quad (8.4)$$

The eigenstates in the $n_h = 1$ sector are the bonding and antibonding molecular orbitals of b_{1g} symmetry with the wavefunctions and energies given by

$$|b,+\rangle = u_b|d_x\rangle + v_b|b_1\rangle, \quad E_{b,+} = \frac{1}{2}(\varepsilon_{dx} + \varepsilon_p - 2t_{pp} + \nu_b),$$
$$|b,-\rangle = -v_b|d_x\rangle + u_b|b_1\rangle, \quad E_{b,-} = \frac{1}{2}(\varepsilon_{dx} + \varepsilon_p - 2t_{pp} - \nu_b). \quad (8.5)$$

Here, the Bogolubov (u,v) coefficients are equal to

$$u_b^2 = \frac{1}{2}\left(1 - \frac{\Delta_b}{\nu_b}\right), \quad v_b^2 = \frac{1}{2}\left(1 + \frac{\Delta_b}{\nu_b}\right), \quad \Delta_b = \delta - 2t_{pp},$$
$$\delta = \varepsilon_p - \varepsilon_{dx}, \quad \nu_b^2 = \Delta_b^2 + 4t_{pd}^2.$$

Similar to the b_{1g} symmetry we can write

$$|a,+\rangle = u_a|d_z\rangle + v_a|a_1\rangle, \quad E_{a,+} = \frac{1}{2}(\varepsilon_{dz} + \varepsilon_p - 2t_{pp} + \nu_a),$$
$$|a,-\rangle = -v_a|d_z\rangle + u_a|a_1\rangle, \quad E_{a,-} = \frac{1}{2}(\varepsilon_{dz} + \varepsilon_p - 2t_{pp} - \nu_a), \quad (8.6)$$

$$u_a^2 = \frac{1}{2}\left(1 - \frac{\Delta_a}{\nu_a}\right), \quad v_a^2 = \frac{1}{2}\left(1 + \frac{\Delta_a}{\nu_a}\right),$$
$$\Delta_a = \varepsilon_p + 2t_{pp} - \varepsilon_{dz}, \quad \nu_a^2 = \Delta_a^2 + 4/3 t_{pd}^2. \quad (8.7)$$

The 3-band p–d model does not take into account the d_z-copper orbital and can be obtained from Eq. (8.4) in the limit $\Delta_d = \varepsilon_{dz} - \varepsilon_{dx} \to \infty$. In this limit in the $n_h = 1$ sector the ground state is given by the molecular orbital $|b,-\rangle$ which corresponds to the covalent admixture of the configurations $d^{10}p^5$ with amplitude u_b and $d^9 p^6$ with amplitude v_b. In the case $t_{pd}/\Delta_b \ll 1$ the admixture is small,

$$u_b^2 \ll 1, \quad v_b^2 \sim 1.$$

The model parameters will be discussed below. For the values typical in cuprates, we have $u_b^2 \approx 0.15$–0.20.

In the two-hole sectors all states are either the spin singlets or the triplets. The Hamiltonian (2.41) matrix for $S = 1, S^z = m, m = 0, \pm 1$ has a size of 15×15 given by the number of various distributions of the 2 holes over 6 atomic orbitals. For the singlet sector the size of the Hamiltonian matrix increases to 21×21 due to the addition of 6 doubly occupied (for the finite values of U_d and U_p) atomic states. The numerical diagonalization of these matrices is carried out with all excited eigenstates taken into account. Of special interest is the triplet–singlet excitation energy $\Delta\varepsilon = \varepsilon_t - \varepsilon_s$ for the singlet and triplet which are the lowest in energy. (Fig. 8.2) (Ovchinnikov 1996). Here, the limit $\Delta_d \to \infty$ corresponds to the three-band p–d model where the d_z states are absent. In this limit for all curves in Fig. 8.2 the value $\Delta\varepsilon$ is rather large, $\Delta\varepsilon \geq 2\,\mathrm{eV}$, in agreement with the three-band model. Nevertheless, the realistic value of Δ_d is not large, $\Delta_d \approx 0.5$–$1\,\mathrm{eV}$, so the limit of small Δ_d is more adequate. It is seen from the curves in Fig. 8.2 that a decreasing Δ_d will result in the decreasing $\Delta\varepsilon$ and the realistic value of $\Delta\varepsilon$ is equal to $\Delta\varepsilon \leq 0.5\,\mathrm{eV}$.

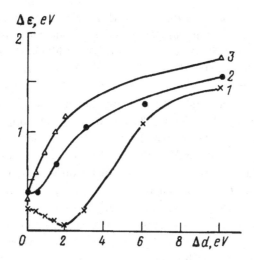

Fig. 8.2 The singlet–triplet excitation energy $\Delta\varepsilon$ as a function of the crystal field splitting parameter Δ_d obtained by the exact diagonalization of the CuO$_4$ cluster at 3 different parameter sets (all in eV): (1) $U_d = 12, U_p = 8, \Delta = 3, t_{pd} = 1.5$, all the others are equal to zero; (2) $V_d = 4.5$, all the others are the same as in (1); (3) $U_d = 12, U_p = 8$, $\Delta = 2, t_{pd} = 1.5, t_{pp} = 0.2; V_d = 4.5; V_p = 3.0; V_{pd} = 0.6, J_p = J_d = 0.5; J_{pd} = 0.6$.

Three parameter sets in Fig. 8.2 corresponds to the following:

(1) Set 1 uses the same parameters as the three-band model plus the crystal field splitting Δ_d, it may be treated as the "minimal set beyond the three-band p–d model".
(2) Set 2 differs from Set 1 by the additional Coulomb interaction of the two holes at different Cu d-orbitals V_d.
(3) Set 3 corresponds to the general case of the multiband p–d model with the matrix elements determined by fitting the optical and magnetic data for the undoped La$_2$CuO$_4$ (Ovchinnikov 1992a).

The small value of the singlet–triplet excitation energy $\Delta\varepsilon$ in the multiband model is in good agreement with the results of Eskes *et al* (1990), Eskes and Sawatzky (1991) and Kamimura and Eto (1991).

There is a simple physical reason as to why the inclusion of the second Cu orbital drastically decreases the stability of the Zhang–Rice singlet. In the three-band model with $d_{x^2-y^2}$ orbital, the virtual copper–oxygen hopping in the second order perturbation process results in the antiferromagnetic interaction similar to the Hubbard model. When the other copper orbital is included the hopping to oxygen and back does not require the singlet state. Moreover, due to the Hund exchange coupling the triplet state gains in energy. Thus, in the multiband model there is a competition of the singlet and triplet states of the two holes while in the three-band p–d model the stability of the Zhang–Rice singlet is overestimated.

The explicit knowledge of all eigenstates with $n_h = 0, 1, 2$ allows us to construct the X-operators and to obtain the exact local Green's function as was discussed in the Chapter 3. The intercluster hopping will be discussed below. Due to the renormalization of the hopping matrix elements $t_{pd} \to \tilde{t}_{pd} = t_{pd}/N^2$, where N is the number of orbitals in the cluster, the dispersion of the quasiparticle bands will be rather small and will reveal itself on the scale $E \sim z\tilde{t}_{pd}$, where z is the nearest neighbors number. For experiments with the low energy resolution, such as X-ray spectroscopy and X-ray photoemission spectroscopy with the resolution $\Delta E \geq 0.5\,\text{eV}$, the band dispersion is less than ΔE and the calculations based on the exact local Green's function with the usual broadening of the delta-function can be used (Avramov and Ovchinnikov 1999).

The single particle density of states (DOS) in the undoped case corresponds to the charge transfer insulator with the gap determined by the charge transfer energy $\delta = \varepsilon_p - \varepsilon_{dx}$. According to the Section 3.4, doping results in some new in-gap states similar to the impurity states in a semiconductor. For p-type doping these states are formed by the mixture of $d_{x^2-y^2}$ copper orbital with the b_1 oxygen orbital with dominant (88%) oxygen contribution. For the n-type doping the in-gap state also occurs (Ovchinnikov 1994) but in this case it is a mixture of d_z (42%) copper and a_1 oxygen (48%) orbitals with a small admixture of the other orbitals.

A sign of the p–d hopping parameter t_{pd} for a given Cu-O pair is determined by the product of the phase factors for d and p wavefunctions. For the chosen phases in the Fig. 8.1, we have

$$t_{pd}(d_{x^2-y^2}, p_{x-1/2}) > 0, \quad t_{pd}(d_{x^2-y^2}, p_{x+1/2}) < 0,$$
$$t_{pd}(d_{x^2-y^2}, p_{y-1/2}) > 0, \quad t_{pd}(d_{x^2-y^2}, p_{y+1/2}) < 0,$$
$$t'_{pd}(d_{z^2}, p_{z-1/2}) < 0, \quad t'_{pd}(d_{z^2}, p_{z+1/2}) > 0.$$

8.2 The Construction of the Wannier Functions and X-Representation of the Multiband p–d Model

The single electron molecular orbitals (8.5) and (8.6) of the isolated CuO_4 cell (and the same for the CuO_6 cell) are not convenient for the description on a lattice because each oxygen ion belongs simultaneously to two adjacent cells that results in the non-orthogonal molecular orbitals at the nearest cells. This requires an explicit determination of the Wannier states. We use the procedure proposed by Zhang and Rice (1988) and Shastry (1989) for the three-band p–d model (which was generalized by Feiner et al (1993) for the multiband model). In this method, with the help of the Fourier transformation

$$d_{\lambda k\sigma} = \sum_f d_{\lambda f\sigma} e^{-ik\cdot f}, \quad p_{\alpha k\sigma} = \sum_m p_{\alpha m\sigma} e^{-ik\cdot m}, \qquad (8.8)$$

the Hamiltonian $H_{pd} \oplus H_{pp}$ can be written in the next form

$$H_{pd} = \frac{t_{pd}}{N} \sum_{k\sigma} d^+_{xk\sigma}(-2i)(s_x p_{xk\sigma} + s_y p_{yk\sigma})$$

$$+ \frac{t_{pd}}{\sqrt{3}N} \sum_{k\sigma} d^+_{zk\sigma} 2i(s_x p_{xk\sigma} - s_y p_{yk\sigma}) + \text{h.c.},$$

$$H_{pp} = -\frac{4t_{pp}}{N} \sum_{k\sigma} s_x s_y p^+_{xk\sigma} p_{yk\sigma} + \text{h.c.}, \quad s_x = \sin(k_x/2), \quad s_y = \sin(k_y/2).$$

$$(8.9)$$

We introduce the new hole operators $a_{k\sigma}$ and $b_{k\sigma}$ by the following linear transformation

$$\begin{pmatrix} b_{k\sigma} \\ a_{k\sigma} \end{pmatrix} = \hat{A} \begin{pmatrix} p_{xk\sigma} \\ p_{yk\sigma} \end{pmatrix} = \begin{pmatrix} is_x/\mu_k & is_y/\mu_k \\ is_y/\mu_k & -is_x/\mu_k \end{pmatrix} \begin{pmatrix} p_{xk\sigma} \\ p_{yk\sigma} \end{pmatrix}$$

where

$$\mu_k^2 = s_x^2 + s_y^2, \quad |\hat{A}| = 1. \qquad (8.10)$$

With the inverse transformation we express the old p-operators via a and b

$$p_{xk\sigma} = -i(s_x b_{k\sigma} + s_y a_{k\sigma})/\mu_k, \quad p_{yk\sigma} = -i(s_y b_{k\sigma} - s_x a_{k\sigma})/\mu_k,$$

and write the Hamiltonian (8.9) in the next form:

$$H_{pd} = -\frac{2t_{pd}}{N} \sum_{k\sigma} \mu_k d^+_{xk\sigma} b_{k\sigma} + \frac{2t_{pd}}{\sqrt{3}N} \sum_{k\sigma} \left(\xi_k d^+_{zk\sigma} b_{k\sigma} + \lambda_k d^+_{zk\sigma} a_{k\sigma} \right) + \text{h.c.}$$

$$= -2t_{pd} \sum_{ij} \mu_{ij} d^+_{xi\sigma} b_{j\sigma} + \frac{2t_{pd}}{\sqrt{3}} \sum_{ij} \left(\xi_{ij} d^+_{zi\sigma} b_{j\sigma} + \lambda_{ij} d^+_{zi\sigma} a_{j\sigma} \right) + \text{h.c.},$$

$$H_{pp} = -\frac{2t_{pp}}{N}\sum_{k\sigma}\nu_k b^+_{k\sigma}b_{k\sigma} + \frac{2t_{pp}}{N}\sum_{k\sigma}\nu_k a^+_{k\sigma}a_{k\sigma} - \frac{2t_{pp}}{N}\sum_{k\sigma}\chi_k\left(b^+_{k\sigma}a_{k\sigma} + \text{h.c.}\right)$$

$$= -2t_{pp}\sum_{ij\sigma}\nu_{ij}b^+_{i\sigma}b_{j\sigma} + 2t_{pp}\sum_{ij\sigma}\nu_{ij}a^+_{i\sigma}a_{j\sigma} + 2t_{pp}\sum_{ij\sigma}\chi_{ij}\left(a^+_{i\sigma}b_{j\sigma} + \text{h.c.}\right),$$

(8.11)

where

$$\lambda_k = 2s_x s_y/\mu_k, \quad \xi_k = (s_x^2 - s_y^2)/\mu_k,$$
$$\nu_k = 4s_x^2 s_y^2/\mu_k^2, \quad \chi_k = 2s_x s_y(s_x^2 - s_y^2)/\mu_k^2.$$

It is easy to verify that the operators a_i, b_i obey the Fermi commutation rules and describe the Wannier states which are orthogonal for different cells

$$\{b_i, b_j^+\} = \frac{1}{N}\sum_{kp}\{b_k, b_p^+\}e^{ikR_i}e^{-ipR_j}$$

$$= \Delta(R_i - R_j) = \begin{cases} 1 & R_i = R_j \\ 0 & R_i \neq R_j \end{cases}.$$

In the Wannier representation both p–d and p–p hoppings become nonlocal. The distance dependence of the matrix elements is given in Table 8.1.

Due to the symmetry $\xi(0,0) = \chi(0,0) = 0$, there is no hybridization of b orbital with a set of a_{1g}-symmetry orbitals, namely the a, d_z, p_z orbitals. A similar procedure for the hopping Hamiltonian of the axial oxygen

$$H'^c_{pp} = t'_{pp}\{p^+_{z+1/2}(p_{x-1/2} - p_{x+1/2} - p_{y-1/2} + p_{y+1/2})$$
$$- p^+_{z-1/2}(p_{x-1/2} - p_{x+1/2} - p_{y-1/2} + p_{y+1/2}) + \text{h.c.}\},$$

$$H'^c_{pd} = -\frac{2t'_{pd}}{\sqrt{3}}\left(p^+_{z+1/2}d_z - p^+_{z-1/2}d_z + \text{h.c.}\right),$$

(8.12)

Table 8.1 The distance dependence of the hopping matrix elements between the cells i and j.

	00	10	11	20	21	22
μ	0.95809	−0.14009	−0.02351	−0.01373	−0.00685	−0.00327
ν	0.72676	−0.27324	0.12207	−0.06385	0.01737	0.01052
λ	0.74587	−0.17578	0.06179	−0.07134	0.01703	0.00925
ξ	0.00000	0.25763	0.00000	0.03913	0.00886	0.00000
χ	0.00000	0.13397	0.00000	−0.04056	0.03043	0.00000

results in the Wannier representation given by

$$H'_{pp} = -\frac{2t'_{pp}}{N} \sum_{k\sigma} \left(\xi_{k\sigma} p^+_{k\sigma} b_{k\sigma} + \lambda_k p^+_{k\sigma} a_{k\sigma} + \text{h.c.} \right)$$

$$= -2t'_{pp} \sum_{ij\sigma} \left(\xi_{ij} p^+_{zi\sigma} b_{j\sigma} + \lambda_{ij} p^+_{zi\sigma} a_{j\sigma} + \text{h.c.} \right), \quad (8.13)$$

$$H'_{pd} = -\frac{2t'_{pd}}{\sqrt{3N}} \sum_{k\sigma} \left(d^+_{zk\sigma} p_{zk\sigma} + \text{h.c.} \right) = -\frac{2t'_{pd}}{\sqrt{3}} \sum_{ij\sigma} \left(d^+_{zi\sigma} p_{zi\sigma} + \text{h.c.} \right).$$

As for the Coulomb p–d and p–p interactions, the three-cell and four-cell terms appear in the Wannier representation (Feiner et al 1993)

$$H^{\text{int}}_{pd} = \sum_{lij} \sum_{\lambda \sigma} V_{pd} \Phi_{lij} n^d_l p^{\lambda+}_{i\sigma} p^\lambda_{j\sigma},$$

$$H^{\text{int}}_{pp} = \sum_{ijkl} \sum_{\lambda_1 \lambda_2 \lambda_3 \lambda_4} U_p \Psi_{ijkl} p^{\lambda_3+}_{i\uparrow} p^{\lambda_4}_{j\uparrow} p^{\lambda_5+}_{k\downarrow} p^{\lambda_6}_{l\downarrow},$$

where $p^{\lambda_i}_{f\sigma} = a_{f\sigma}, b_{f\sigma}$.
The largest coefficients Φ_{lij} and Ψ_{klij} are

$$\Phi_{000} = 0.918; \quad \Phi_{001} = -0.13; \quad \Phi_{002} = -0.02;$$

$$\Psi_{0000} = 0.2109; \quad \Psi_{0001} = -0.03.$$

Due to the small values of all intercell coefficients we only take into account the intracell p–d and p–p interactions

$$H^{\text{int}}_{pd} = V_{pd} \Phi_{000} \sum_{i\lambda\sigma\sigma'} n^d_{i\sigma} n^\lambda_{i\sigma'}, \quad H^{\text{int}}_{pp} = U_p \Psi_{0000} \sum_i n^p_{i\uparrow} n^p_{i\downarrow},$$

where $n^d_{i\sigma} = \sum_\lambda d^+_{\lambda i\sigma} d_{\lambda i\sigma}$; $n^p_{i\sigma} = \sum_\lambda p^{\lambda+}_{i\sigma} p^\lambda_{i\sigma}$.
The total multiband p–d model Hamiltonian in the Wannier representation is given by the sum of the intracell and the intercell parts:

$$H = H_c + H_{cc}, \quad H_c = \sum_f H_f, \quad H_f = h^{(b)} + h^{(a)} + h^{(ab)}, \quad (8.14)$$

$$h^{(b)} = \sum_\sigma \left(\varepsilon_b n^b_\sigma + \varepsilon_{d_x} n^{d_x}_\sigma \right) + U_d n^{d_x}_\uparrow n^{d_x}_\downarrow + U_b n^b_\uparrow n^b_\downarrow$$

$$+ V_{pd} n^{d_x} n^b - \tau_b \sum_\sigma \left(d^+_{x\sigma} b_\sigma + \text{h.c.} \right),$$

$$h^{(a)} = \sum_\sigma \left(\varepsilon_a n^a_\sigma + \varepsilon_{d_z} n^{d_z}_\sigma + \varepsilon_{p_z} n^{p_z}_\sigma\right) + U_d n^{d_z}_\uparrow n^{d_z}_\downarrow + U_a n^a_\uparrow n^a_\downarrow$$

$$+ V'_{pd} n^{d_z} n^{p_z} + U'_p n^{p_z} n^{p_z} + V_{pd} n^{d_z} n^a + \tau_a \sum_\sigma \left(d^+_{z\sigma} a_\sigma + \text{h.c}\right)$$

$$- \tau'_{pd} \sum_\sigma \left(d^+_{z\sigma} p_{z\sigma} + \text{h.c.}\right) - t'_{pp} \sum_\sigma \left(a^+_\sigma p_{z\sigma} + \text{h.c}\right);$$

$$h^{(ab)} = U_d n^{d_x} n^{d_z} + U_{ab} n^a n^b + V_{pd} n^{d_x} n^a + V_{pd} n^b n^{d_z} + V'_{pd} n^{d_x} n^{p_z},$$

$$H_{cc} = \sum_{ij} \sum_\sigma \left(h^{(b)}_{hop} + h^{(a)}_{hop} + h^{(ab)}_{hop}\right),$$

$$h^{(b)}_{hop} = -2t_{pd}\mu_{ij} d^+_{xi\sigma} b_{j\sigma} - 2t_{pp}\nu_{ij} b^+_{i\sigma} b_{j\sigma} + \text{h.c.},$$

$$h^{(a)}_{hop} = \frac{2t_{pd}}{\sqrt{3}} \lambda_{ij} d^+_{zi\sigma} a_{j\sigma} + 2t_{pp}\nu_{ij} a^+_{i\sigma} a_{j\sigma} - 2t'_{pp} \lambda_{ij} p^+_{zi\sigma} a_{j\sigma} + \text{h.c.},$$

$$h^{(ab)}_{hop} = \frac{2t_{pd}}{\sqrt{3}} \xi_{ij} d^+_{zi\sigma} b_{j\sigma} + 2t_{pp}\chi_{ij} a^+_{i\sigma} b_{j\sigma} - 2t'_{pp}\xi_{ij} p^+_{zi\sigma} b_{j\sigma} + \text{h.c.},$$

where

$$\varepsilon_b = \varepsilon_p - 2t_{pp}\nu_{00}, \quad \varepsilon_a = \varepsilon_p + 2t_{pp}\nu_{00},$$

$$\tau_b = 2t_{pd}\mu_{00}, \quad \tau_a = 2t_{pd}\lambda_{00}/\sqrt{3},$$

$$\tau'_{pd} = 2t'_{pd}/\sqrt{3}, \quad \tau'_{pp} = 2t'_{pp}\lambda_{00}.$$

Here, for simplicity all p–p Coulomb intracell matrix elements are taken to be equal:

$$U_a = U_{ab} = U'_p = U_b = U_p \Psi_{0000}, \quad V_{pd} = V_{pd}\Phi_{000}.$$

The next step in our approach is the exact diagonalization of the intracell Hamiltonian H_c. As before we do it separately in the Hilbert subspace with the number of holes $n_h = 0, 1, 2$. In the vacuum sector $n_h = 0$ there is only one state $|0\rangle$ with energy equal to zero. In the $n_h = 1$ sector the H_c has a block structure corresponding to the a_{1g} and b_{1g} molecular orbitals. The b_{1g} eigenvectors are given by

$$|\tilde{b}_{i\sigma}\rangle = \beta_i(b) b^+_\sigma |0\rangle + \beta_i(d_x) d^+_{i\sigma} |0\rangle$$

with the energies $\varepsilon_{1,\tilde{b}_i}$ and the indexes $i = 1, 2$ corresponding to the bonding and antibonding states. They are obtained by the diagonalization of the following matrix

$$\hat{h}^{(b)} = \begin{pmatrix} \varepsilon_{d_x} & -\tau_b \\ -\tau_b & \varepsilon_b \end{pmatrix}. \quad (8.15)$$

The a_{1g} eigenstate is the mixture of a, p_z and d_z orbitals

$$|\tilde{a}_{i\sigma}\rangle = \alpha_i(a)a_\sigma^+|0\rangle + \alpha_i(p_z)p_{z\sigma}^+|0\rangle + \alpha_i(d_z)d_{z\sigma}^+|0\rangle$$

with the energies ε_{1,a_i} ($i = 1, 2, 3$) determined by the diagonalization of the matrix

$$\hat{h}^{(a)} = \begin{pmatrix} \varepsilon_{d_z} & \tau_a & -\tau'_{pd} \\ \tau_a & \varepsilon_a & -t'_{pp} \\ -\tau'_{pd} & -t'_{pp} & \varepsilon_{p_z} \end{pmatrix}. \quad (8.16)$$

In the two-hole sector, one has $^1A_1$ singlets and $^3B_1$ triplets. The lowest and excited singlets are written in the form $|\tilde{A}_p\rangle = \sum_i A_{qi}|A_i\rangle$ with $i, q = 1$–9. The coefficients A_{qi} and vectors $|A_i\rangle$ given in the Table 8.2. The eigenenergies $\varepsilon_{2\tilde{A}_q}$ are obtained by the exact diagonalization of the following matrix

$$\hat{h}^{(A)} = \begin{pmatrix} \hat{h}_{11}^{(A)} & 0 \\ 0 & \hat{h}_{22}^{(A)} \end{pmatrix}, \quad (8.17)$$

Table 8.2 The basis vectors and their amplitudes in the singlet two-hole sector for the CuO_6 cluster. Here, $|ZR\rangle$ is the Zhang–Rice singlet.

i	A_{qi}	$|A_i\rangle$
1	$A_q(d_x b)$	$|ZR\rangle = \frac{1}{\sqrt{2}}|d_{x\downarrow}^+ b_\uparrow^+ - d_{x\uparrow}^+ b_\downarrow^+|0\rangle$
2	$A_q(bb)$	$|b_\downarrow^+ b_\uparrow^+|0\rangle$
3	$A_q(d_x d_x)$	$|d_{x\downarrow}^+ d_{x\uparrow}^+|0\rangle$
4	$A_q(p_z a)$	$\frac{1}{\sqrt{2}}|p_{z\downarrow}^+ a_\uparrow^+ - p_{z\uparrow}^+ a_\downarrow^+|0\rangle$
5	$A_q(d_z a)$	$\frac{1}{\sqrt{2}}|d_{z\downarrow}^+ a_\uparrow^+ - d_{z\uparrow}^+ a_\downarrow^+|0\rangle$
6	$A_q(d_z p_z)$	$\frac{1}{\sqrt{2}}|d_{z\downarrow}^+ p_{z\uparrow}^+ - d_{z\uparrow}^+ p_{z\downarrow}^+|0\rangle$
7	$A_q(aa)$	$|a_\downarrow^+ a_\uparrow^+|0\rangle$
8	$A_q(p_z p_z)$	$|p_{z\downarrow}^+ p_{z\uparrow}^+|0\rangle$
9	$A_q(d_z d_z)$	$|d_{z\downarrow}^+ d_{z\uparrow}^+|0\rangle$

where

$$\hat{h}_{11}^{(A)} = \begin{pmatrix} \varepsilon_b + \varepsilon_{d_x} + V_{pd} & -\sqrt{2}\tau_b & \sqrt{2}\tau_b \\ -\sqrt{2}\tau_b & 2\varepsilon_b + U_b & 0 \\ -\sqrt{2}\tau_b & 0 & 2\varepsilon_{d_x} + U_d \end{pmatrix},$$

$$\hat{h}_{22}^{(A)}$$

$$= \begin{pmatrix} \varepsilon_a + \varepsilon_{p_z} + V'_p & -\tau'_{pdf} & \tau_a & -\sqrt{2}t'_{pp} & -\sqrt{2}t'_{pp} & 0 \\ -\tau'_{pdf} & \varepsilon_{d_z} + \varepsilon_a + V_{pd} & -t'_{pp} & \sqrt{2}\tau_a & 0 & \sqrt{2}\tau_a \\ \tau_a & -t'_{pp} & \varepsilon_{d_z} + \varepsilon_{p_z} + V'_{pd} & 0 & -\sqrt{2}\tau'_{pdf} & -\sqrt{2}\tau'_{pdf} \\ -\sqrt{2}t'_{pp} & \sqrt{2}\tau_a & 0 & 2\varepsilon_a + U_a & 0 & 0 \\ -\sqrt{2}t'_{pp} & 0 & -\sqrt{2}\tau'_{pdf} & 0 & 2\varepsilon_{p_z} + U'_p & 0 \\ 0 & \sqrt{2}\tau_a & -\sqrt{2}\tau'_{pdf} & 0 & 0 & 2\varepsilon_{d_z} + U_d \end{pmatrix}.$$

In the triplet two-hole subspace, the eigenvectors are in the form $|\tilde{B}_{Mq}\rangle = \sum_i B_{qi}|B_{iM}\rangle$ with $i,q = 1\text{-}6$, $M = -1, 0, 1$ and the basis vectors given in the Table 8.3. The triplet state energies can be obtained from the diagonalization of the following matrix

$$\hat{h}^{(\tilde{B})}$$

$$= \begin{pmatrix} \varepsilon_a + \varepsilon_{d_x} + V_{pd} & -\tau_b & \tau_a & 0 & -t'_{pp} & 0 \\ -\tau_b & \varepsilon_a + \varepsilon_b + U_b & 0 & \tau_a & 0 & -t'_{pp} \\ \tau_a & 0 & \varepsilon_{d_z} + \varepsilon_{d_x} + U_d & -\tau_b & -\tau'_{pd} & 0 \\ 0 & \tau_a & -\tau_b & \varepsilon_{d_z} + \varepsilon_b + V_{pd} & 0 & -\tau'_{pd} \\ -t'_{pp} & 0 & -\tau'_{pd} & 0 & \varepsilon_{d_x} + \varepsilon_{p_z} + V'_{pd} & -\tau_b \\ 0 & -t'_{pp} & 0 & -\tau'_{pd} & -\tau_b & \varepsilon_b + \varepsilon_{p_z} + V'_{pp} \end{pmatrix}$$

(8.18)

Table 8.3 The basis vectors and their amplitudes in the triplet two-hole sector for the CuO_6 cluster.

I	B_{qi}	$|B_{i-1}\rangle$	$|B_{i0}\rangle$	$|B_{i1}\rangle$
1	$B_q(d_x a)$	$|d^+_{x\downarrow}a^+_\downarrow|0\rangle$	$\frac{1}{\sqrt{2}}|d^+_{x\downarrow}a^+_\uparrow + d^+_{x\uparrow}a^+_\downarrow|0\rangle$	$|d^+_{x\uparrow}a^+_\uparrow|0\rangle$
2	$B_q(ba)$	$|b^+_\downarrow a^+_\downarrow|0\rangle$	$\frac{1}{\sqrt{2}}|b^+_\downarrow a^+_\uparrow + b^+_\uparrow a^+_\downarrow|0\rangle$	$|b^+_\uparrow a^+_\uparrow|0\rangle$
3	$B_q(d_x d_z)$	$|d^+_{x\downarrow}d^+_{z\downarrow}|0\rangle$	$\frac{1}{\sqrt{2}}|d^+_{x\downarrow}d^+_{z\uparrow} + d^+_{x\uparrow}d^+_{z\downarrow}|0\rangle$	$|d^+_{x\uparrow}d^+_{z\uparrow}|0\rangle$
4	$B_q(d_z b)$	$|d^+_{z\downarrow}b^+_\downarrow|0\rangle$	$\frac{1}{\sqrt{2}}|d^+_{z\downarrow}b^+_\uparrow + d^+_{z\uparrow}b^+_\downarrow|0\rangle$	$|d^+_{z\uparrow}b^+_\uparrow|0\rangle$
5	$B_q(d_x p_z)$	$|d^+_{x\downarrow}p^+_{z\downarrow}|0\rangle$	$\frac{1}{\sqrt{2}}|d^+_{x\downarrow}p^+_{z\uparrow} + d^+_{x\uparrow}p^+_{z\downarrow}|0\rangle$	$|d^+_{x\uparrow}p^+_{z\uparrow}|0\rangle$
6	$B_q(b p_z)$	$|b^+_\downarrow p^+_{z\downarrow}|0\rangle$	$\frac{1}{\sqrt{2}}|b^+_\downarrow p^+_{z\uparrow} + b^+_\uparrow p^+_{z\downarrow}|0\rangle$	$|b^+_\uparrow p^+_{z\uparrow}|0\rangle$

The Electronic Structure of Copper Oxides in the Multiorbital p–d Model   207

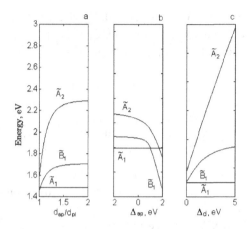

Fig. 8.3 The dependence of the lowest in energy singlet $^1A_1, {}^1A_2$ and triplet $^3B_1$ terms of the CuO$_6$ cluster on the following parameters: (a) The ratio of the (Cu-axial O)/(Cu-plane O) distances with $\Delta_d = \varepsilon_{d_z} - \varepsilon_{d_x} = 2\,\text{eV}, \Delta_{ap} = \varepsilon_p - \varepsilon_{p_z} = 0.5\,\text{eV}$. (b) The oxygen crystal field splitting, Δ_{ap} with $d_{ap}/d_{pl} = 1.2, \Delta_d = 2\,\text{eV}$. (c) The copper crystal field splitting Δ_d with $d_{ap}/d_{pl} = 1.2, \Delta_{ap} = 0.5\,\text{eV}$.

The low energy singlets $|\tilde{A}_1\rangle, |\tilde{A}_2\rangle$ and triplet $^3B_1$ are shown in Fig. 8.3 as functions of the model parameters. The lowest singlet is formed by the mixture of the three configurations

$$|\tilde{A}_1\rangle = A_1(d_x b)|ZR\rangle + A_2(bb)|A_2\rangle + A_3(d_x d_x)|A_3\rangle$$

For the realistic values of the parameters (see below) the amplitude of Zhang–Rice singlet is dominant. For the triplet term the different configurations contribute differently depending on the parameters and there is no dominant configuration. With decreasing energy Δ_{ap} in the axial oxygen the amplitude of the $d_x p_z$ configuration B_5 increases and the triplet–singlet crossover is possible (Fig. 8.3(b)). Decreasing the copper crystal field parameter Δ_d results in an increase in the amplitude of the Hund's configuration $d_x d_z B_3$ and a decrease in the triplet–singlet excitation energy $\Delta\varepsilon$ (Fig. 8.3(c)). As for the CuO$_4$ cluster, the exact diagonalization of the multiband p–d Hamiltonian for the CuO$_6$ cluster gives $\Delta\varepsilon = 0.5\,\text{eV}$ for realistic parameter values. It means that contrary to the one-band Hubbard model the realistic model should take into account both the singlet and the triplet two-hole states.

After the exact diagonalization of the cell Hamiltonian H_c we construct the cell X-operators in the usual way, $X_f^{pp} = |p\rangle\langle q|$ where the states $\{|p\rangle\}$ include all eigenstates with $n_h = 0, 1, 2$. To consider the antiferromagnetic state we introduce two sublattices, A and B, with small effective staggered field $h \sim k_B T_N$ splitting the spin sublevels $\varepsilon_{1pA} = \varepsilon_{1p} - \sigma h$, $\varepsilon_{1pB} = \varepsilon_{1p} + \sigma h$. The cell Hamiltonian in the X-operator representation is written in the form

$$H_c = \sum_{p\sigma}(\varepsilon_{1pG} - \mu)X^{pp}_{f_G\sigma} + \sum_{qf_G}(\varepsilon_{2qG} - 2\mu)X^{qq}_{f_G\sigma},$$

$$\text{where } f_G = \begin{cases} f_A, & f \in A \\ f_B, & f \in B \end{cases}. \quad (8.19)$$

In the X-operator representation the annihilation and creation operators for the electron at the cell f, orbital λ and spin projection σ are given by

$$a_{f\lambda\sigma} = \sum_m \gamma_{\lambda\sigma}(m)X^m_f, \quad a^+_{f\lambda\sigma} = \sum_m \gamma^*_{\lambda\sigma}(m)X^{+m}_f$$

where $c_{f\lambda\sigma} = \{d_{fx\sigma}, d_{fz\sigma}, a_{f\sigma}, b_{f\sigma}, p_{fz\sigma}\}$ and m is the number of the quasiparticles, $m \leftrightarrow \alpha_m(p,q)$. The matrix elements $\gamma_{\lambda\sigma}(m)$ are calculated straightforwardly. If we restrict ourselves to the ground and the lowest excited term in both the $n_h = 1$ and $n_h = 2$ sectors, the number of different Fermi type quasiparticles is equal to $8 \times 2 = 32$[(spin projection)×(two sublattices)]. All other excited single- and two-hole states are not essential in the low energy physics. It means that we include only two $|p\rangle (= |\tilde{a}_1\rangle$ and $|\tilde{b}_1\rangle)$ states and two $|q\rangle (= |\tilde{A}_1\rangle$ and $|\tilde{B}_{1M}\rangle)$ states in Eq. (8.19) as well as the vacuum state $|0\rangle$. The matrix elements for the excitations $\tilde{b}_1 \to |\tilde{A}_1\rangle$, $\tilde{a}_1 \to |\tilde{B}_{1M}\rangle$ and $\tilde{b}_1 \to |\tilde{B}_{1M}\rangle$ are given in Tables 8.4, 8.5 and 8.6 respectively.

The intercell hopping between the intrasublattice neighbors along the distance R_1 and for the intersublattice hopping along R_2 may be written

Table 8.4 The matrix elements for the quasiparticles α_m ($\tilde{b}_1, \tilde{A}_1$) with $c_i = d_x b$. For all the other values of m, the matrix elements are equal to zero.

f_G	f_A		f_B	
σ'	↑	↓	↑	↓
m	1	9	17	25
$\gamma_{d_x\sigma}$	$\eta(\sigma)(\delta_{\sigma\sigma'} - 1)\sum_i \beta_1(c_i)A_1(d_x c_i)$			
$\gamma_{b\sigma}$	$\eta(\sigma)(\delta_{\sigma\sigma'} - 1)\sum_i \beta_1(c_i)A_1(bc_i)$			

Table 8.5 The matrix elements for the quasiparticles $\alpha_m(\tilde{a}_{1\sigma'}\tilde{B}_{1M})$ with $c_i = d_z, a, p_z$.

α_m	$\alpha_m(\tilde{a}_{1\sigma'}\tilde{B}_{1-1})$			$\alpha_m(\tilde{a}_{1\sigma'}\tilde{B}_{10})$				$\alpha_m(\tilde{a}_{1\sigma'}\tilde{B}_{11})$			
f_G	f_A		f_B		f_A		f_B		f_A		f_B	
σ'	↑	↓	↑	↓	↑	↓	↑	↓	↑	↓	↑	↓
m	2	10	18	26	3	11	19	27	4	12	20	27
$\gamma_{d_x\sigma}$	$\delta_{\sigma\sigma'}\sum_i \alpha_1(c_i)B_1(d_xc_i)$				$\frac{1}{\sqrt{2}}(1-\delta_{\sigma\sigma'})\sum_i \alpha_1(c_i)B_1(d_xc_i)$				$\delta_{\sigma\sigma'}\sum_i \alpha_1(c_i)B_1(d_xc_i)$			
$\gamma_{b\sigma}$	$\delta_{\sigma\sigma'}\sum_i \alpha_1(c_i)B_1(bc_i)$				$\frac{1}{\sqrt{2}}(1-\delta_{\sigma\sigma'})\sum_i \alpha_1(c_i)B_1(bc_i)$				$\delta_{\sigma\sigma'}\sum_i \alpha_1(c_i)B_1(bc_i)$			

Table 8.6 The matrix elements for the quasiparticles $\alpha_m(\tilde{b}_{1\sigma'}\tilde{B}_{1M})$ with $M = -1, 0, +1, c_i = d_x, b$.

α_m	$\alpha_m(\tilde{b}_{1\sigma'}\tilde{B}_{1-1})$				$\alpha_m(\tilde{b}_{1\sigma'}\tilde{B}_{10})$				$\alpha_m(\tilde{b}_{1\sigma'}\tilde{B}_{11})$			
f_G	f_A		f_B		f_A		f_B		f_A		f_B	
σ'	↑	↓	↑	↓	↑	↓	↑	↓	↑	↓	↑	↓
M	5	13	21	28	6	14	22	29	7	15	23	21
$\gamma_{a\sigma}$	$-\delta_{\sigma\sigma'}\sum_i \beta_1(c_i)B_1(ac_i)$				$-\frac{1}{\sqrt{2}}(1-\delta_{\sigma\sigma'})\sum_i \beta_1(c_i)B_1(ac_i)$				$-\delta_{\sigma\sigma'}\sum_i \beta_1(c_i)B_1(ac_i)$			
$\gamma_{d_z\sigma}$	$-\delta_{\sigma\sigma'}\sum_i \beta_1(c_i)B_1(p_zc_i)$				$-\frac{1}{\sqrt{2}}(1-\delta_{\sigma\sigma'})\sum_i \beta_1(c_i)B_1(p_zc_i)$				$-\delta_{\sigma\sigma'}\sum_i \beta_1(c_i)B_1(p_zc_i)$			
$\gamma_{p_z\sigma}$	$-\delta_{\sigma\sigma'}\sum_i \beta_1(c_i)B_1(d_zc_i)$				$-\frac{1}{\sqrt{2}}(1-\delta_{\sigma\sigma'})\sum_i \beta_1(c_i)B_1(d_zc_i)$				$-\delta_{\sigma\sigma'}\sum_i \beta_1(c_i)B_1(d_zc_i)$			

in the antiferromagnetic state as

$$H_{cc} = \begin{pmatrix} H_{AA} & H_{AB} \\ H_{BA} & H_{BB} \end{pmatrix}$$

$$= \sum_{\lambda\lambda'\sigma} \begin{pmatrix} \sum_{fR_1} T_{\lambda\lambda'}(R_1)(c^+_{f\lambda\sigma}c_{f+R_1\lambda'\sigma} + \text{h.c.}) & \sum_{fR_2} T_{\lambda\lambda'}(R_2)(c^+_{f\lambda\sigma}c_{f+R_1\lambda'\sigma} + \text{h.c.}) \\ \sum_{gR_2} T_{\lambda\lambda'}(R_2)(c^+_{g\lambda\sigma}c_{g+R_2\lambda'\sigma} + \text{h.c.}) & \sum_{gR_1} T_{\lambda\lambda'}(R_1)(c^+_{g\lambda\sigma}c_{g+R_1\lambda'\sigma} + \text{h.c.}) \end{pmatrix}$$

$$= \sum_{\lambda\lambda'\sigma}\sum_{kmn} \gamma^*_{\lambda\sigma}(m)\gamma_{\lambda'\sigma}(n) \begin{pmatrix} T^{AA}_{\lambda\lambda'}(k)X^{+m}_{k\sigma}X^n_{k\sigma} & T^{AB}_{\lambda\lambda'}(k)X^{+m}_{k\sigma}Y^n_{k\sigma} \\ T^{BA}_{\lambda\lambda'}(k)Y^{+m}_{k\sigma}X^n_{k\sigma} & T^{BB}_{\lambda\lambda'}(k)Y^{+m}_{k\sigma}Y^n_{k\sigma} \end{pmatrix} + \text{h.c.}$$

(8.20)

where

$$T^{AA}_{\lambda\lambda'}(k) = \frac{2}{N}\sum_R T^{AA}_{\lambda\lambda'}(R_1)e^{ikR_1} = T^{BB}_{\lambda\lambda'}(k),$$

$$T^{AB}_{\lambda\lambda'}(k) = \frac{2}{N}\sum T^{AB}_{\lambda\lambda'}(R_2)e^{ikR_2} = T^{BA}_{\lambda\lambda'}(k).$$

Here $X_{k\sigma}^m, Y_{k\sigma}^n$ are the Fourier transforms of the Hubbard operators in the A and B sublattice. Under the basis states of d_x, d_z, b, a and p_z orbitals, the intracell hopping matrix is given by

$$T_{\lambda\lambda'}(\boldsymbol{R}) = \begin{pmatrix} 0 & 0 & -2t_{pd}\mu & 0 & 0 \\ 0 & 0 & 2t_{pd}\xi/\sqrt{3} & 2t_{pd}\lambda/\sqrt{3} & 0 \\ -2t_{pd}\mu & 2t_{pd}\xi/\sqrt{3} & -2t_{pp}\nu & 2t_{pp}\chi & -2t'_{pp}\xi \\ 0 & 2t_{pd}\lambda/\sqrt{3} & 2t_{pp}\chi & 2t_{pp}\nu & -2t'_{pp}\lambda \\ 0 & 0 & -2t'_{pp}\xi & -2t'_{pp}\lambda & 0 \end{pmatrix} \quad (8.21)$$

According to the approach of Chapter 3 the quasiparticle band structure is determined by the following dispersion equation of the generalized tight-binding method

$$\left\| \frac{(E - \Omega_m^G)\delta_{mn}}{F_\sigma^G(m)} - 2\sum_{\lambda\lambda'} \gamma_{\lambda\sigma}^*(m) T_{\lambda\lambda'}^{FG}(\boldsymbol{k}) \gamma_{\lambda'\sigma}(n) \right\| = 0 \quad (8.22)$$

where $F, G = A, B$ are the sublattice indexes.

8.3 The Evolution of the Band Structure and ARPES Data with Doping from the Undoped Antiferromagnetic Insulator to Optimal Doped and Overdoped Metal

The band structure for the undoped case $n_h = 1$ has been calculated (Gavrichkov et al 2000) for the following set of the model parameters (in t_{pd} units)

$$\begin{aligned} \varepsilon_{d_x} &= 0; \quad \varepsilon_{d_z} = 2; \quad \varepsilon_p = 1.6; \quad \varepsilon_{p_R} = 0.5; \quad t_{pp} = 0.46; \quad t'_{pp} = 0.42; \\ U_d &= 9; \quad U_p = 4; \quad V_{pd} = 1.5; \quad J_d = 1. \end{aligned} \quad (8.23)$$

The top of the valence band is shown in Fig. 8.4 for the four symmetrical lines in the Brillouin zone and is compared with the angle resolved photoemission electron spectra (ARPES) for the undoped antiferromagnet compound $Sr_2CuO_2Cl_2$ (Wells et al 1995). The empty conductivity band is separated from the valence band by the charge-transfer gap $E_g \approx 2\,\text{eV}$. Several other subbands in the valence band are below the energy scale of Fig. 8.4. These subbands are formed by the quasiparticle $\alpha_m(\tilde{b}_{1\sigma}\tilde{B}_{1M})$ with the participation of two-hole triplet states. Its dispersion is less than the top of the valence band due to zero matrix elements $\gamma_{d_x}(m), \gamma_b(m)$ for the

The Electronic Structure of Copper Oxides in the Multiorbital p–d Model   211

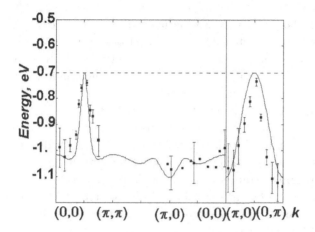

Fig. 8.4 The dispersion of the top of the valence band (solid line) calculated by the generalized tight-binding method and the experimental data by ARPES for $Sr_2CuO_2Cl_2$ (Wells et al 1995). The dotted line at the top of the band shows the in-gap state with zero spectral weight in the undoped case.

corresponding numbers m. The top of the valence band is formed mainly by the quasiparticle $\alpha_m(\tilde{b}_{1\sigma}\tilde{A}_1)$ (singlet excitation) with the significant contribution from $\alpha_m(\tilde{b}_{1\sigma}\tilde{B}_{1M})$ (triplet excitation). It is clear from Fig. 8.4 that our band structure is in good agreement with the ARPES data for the undoped $Sr_2CuO_2Cl_2$. The t–J model fits the ARPES data only in the $[1, 1]$ direction of the Brillouin zone, while the dispersion in the $[1, -1]$ direction is too small and the hole energy along the $[0, 1]$ direction is too high. The addition of the intrasublattice hopping t' and t'' in the t–J model provides better agreement in the $[1, -1]$ direction (Daffy et al 1997). Nevertheless there is still some significant discrepancy in the $[0, 1]$ direction. Moreover, the experimental linewidth of the ARPES spectra in the $[0, 1]$ direction is very large (see Fig. 8.4).

In our approach these facts can simply be explained by the contribution of the triplet excitation which is absent in the t–J model. As mentioned above in the multiband p–d model the triplet two-hole state is not high in energy above the singlet state with the excitation energy $\Delta\varepsilon \approx 0.5$ eV. In the intercell hopping Hamiltonian H_{cc} there are nonzero matrix elements providing the mixture of the $\alpha(\tilde{b}_{1\sigma}\tilde{A}_1)$ and $\alpha(\tilde{b}_{1\sigma}\tilde{B}_1)$ quasiparticles. This mixture has the maximum effect near the $X = (\pi, 0)$ point, decreasing the hole energy in comparison with t–J and t'–t–J model in the range of the ARPES data. The multiband singlet–triplet transitions result in the additional scattering of the hole and increase the linewidth of the ARPES experiments near the $(\pi, 0)$ wavevector. The other effect of the triplet admixture is the dispersion along the $[1, -1]$ direction $(0, \pi)$–$(\pi, 0)$ which is the Brillouin zone boundary in the antiferromagnetic phase and where the intersublattice hopping cannot occur. Nevertheless, the bandwidth in this direction is almost equal to the bandwidth in the $[1, 1]$ direction (Wells et al 1995; see Fig. 8.4). In the t–J model this dispersion appears due to the next neighboring hopping t' while in our approach the bandwidth in the $[1, -1]$ direction is determined by mixing the singlet and triplet excitations. When ignoring the triplet state we have a small dispersion in the $[1, -1]$ direction similar to the t–J model (Liu and Manousakis 1992), the bandwidth of the valence band in this case is larger than the experimental one; this is similar to LDA calculations by Andersen et al (1995). The effective t and t' parameters in our approach that have the same meaning as in the t–t'–J model are given by

$$\frac{t'_{eff}}{t_{eff}} = \frac{t_{pd}\mu_{11}\varphi_d\varphi_p + t_{pp}\nu_{11}\varphi_p^2}{t_{pd}\mu_{10}\varphi_d\varphi_p + t_{pp}\nu_{10}\varphi_p^2} = -0.24, \qquad (8.24)$$

with

$$\varphi_d = \beta_1(d_x)A_1(d_xd_x) + \beta_1(b)A_1(d_xb),$$
$$\varphi_b = \beta_1(d_x)A_1(d_xb) + \beta_1(b)A_1(bb).$$

The ratio (8.24) is quite consistent with the value $t'/t = -0.35$ obtained by Daffy et al (1997) via a fitting to the ARPES data.

The dotted line at the top of the valence band shown in Fig. 8.4 corresponds to the in-gap impurity-like state formed by the excitation of the spin-minority $|\tilde{b}_{1,-\sigma}\rangle$ state for the given sublattice ($\sigma = +1/2$ for $f \in A$, $\sigma = -1/2$ for $f \in B$) to the two-hole state $|\tilde{A}_1\rangle$. At zero temperature, neglecting the quantum spin fluctuations in the undoped case $x = 0$, the filling factor for the in-gap state $\langle X^{\tilde{b}_{1,-\sigma},\tilde{b}_{1,-\sigma}}\rangle + \langle X^{\tilde{A}_1,\tilde{A}_1}\rangle = 0$ gives zero spectral weight and zero dispersion of this state. Therefore, we call this state a virtual level. But with doping the filling factor $n_{\tilde{A}_1} = \langle X^{\tilde{A}_1,\tilde{A}_1}\rangle = x$ and the virtual level acquires a dispersion and a spectral weight proportional to x. The evolution of the quasiparticle band structure with hole doping is shown in Fig. 8.5. The dispersion of the former virtual level into narrow band is the main effect of doping on the electronic structure. This band looks like an impurity band in semiconductors because its spectral weight

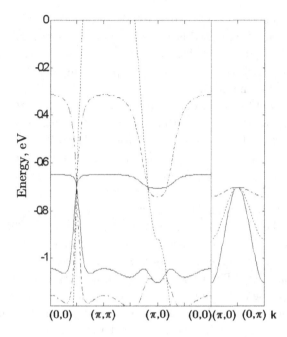

Fig. 8.5 The dispersion of the valence band of the CuO_2 layer doped by holes, $x = 0.01$ and $x = 0.10$ in the antiferromagnetic phase are shown by solid and dot-dashed lines. The dotted line corresponds to the paramagnetic state with $x = 0.20$ (Gavrichkov et al 2000).

is proportional to x. It gives in the in-gap state near the top of the valence band in the single particle DOS. It should be emphasized that the impurity band in the SEC systems forms due to mechanism of the spectral weight redistribution (Ovchinnikov 1992b) with a simultaneous decrease $\sim(1-x)$ of the spectral weight of the empty conductivity band and triplet excitation $\alpha(\tilde{b}_{1\sigma}\tilde{B}_1)$ in the valence band. There are no fluctuations of the crystal potential or hopping integrals which result in the impurity state splitting in the conventional single electron band theory.

The main change in the dispersion of the valence band with doping is the appearance of the maximum at the momentum $\bar{M} \equiv \left(\frac{\pi}{a}, \frac{\pi}{a}\right)$ instead of $\bar{M} \equiv \left(\frac{\pi}{2a}, \frac{\pi}{2a}\right)$ in the undoped case and the minimum at $(\pi, 0) \equiv X$ point. The application of our approach to the region above x_{cr} where long-range antiferromagnetic order is absent requires a special discussion. In principle, the self-consistent treatment of the electron structure and magnetic order is possible; for more simple case of the t–J model a similar approach has been realised by Plakida and Oudovenko (1999). In the mean field approximation of the "Hubbard 1" type used here the magnetic subsystem is described by the effective Heisenberg Hamiltonian and it influences the electronic structure only via the redistribution of the occupation numbers between the spin split sublevels for the given sublattice. The results of this approximation has been analyzed by Ovchinnikov (1995) and compared with the results of the exact diagonalization (ED) study of the Hubbard model, t–J model and three-band p–d model. Thus, the comparison of the $J/t \ll 1$ perturbation theory with the ED of the t–J model reveals that the vertex renormalization is small (Martinez and Horsch 1991). For the undoped Hubbard model the quasiparticle spectral density obtained by the ED method can be described by the mean field theory with the sublattice magnetization calculated in the spin wave approximation (Dagotto et al 1992). The extra hole induced by doping occupies mainly the oxygen orbitals and results in an additional exchange interaction $J_{Cu-O} \gg J_{Cu-Cu}$ (Guo et al 1988; Aharony et al 1988). The three dimensional antiferromagnetic long-range order is suppressed by doping, the critical concentration x_{cr} (Khaliulin 1990; Khaliulin and Horsch 1993) and the concentration dependence of the Neel temperature $T_N(x)$ (Richard and Yushankhai 1994) have been calculated via the spin-wave approach in the framework of the t–J model. The one-hole quasiparticle in the two-dimensional doped Mott–Hubbard antiferromagnet is known to be a spin polaron (see Nagaev 1983; Barabanov et al 1989; Barabanov et al 1991; Hayn et al 1996; also see reviews by Dagotto 1994; Kampf 1994; Brenig 1995; Izyumov 1995). While the long-range antiferromagnetic order above x_{cr} is absent, there exists a short range order with a reasonably large correlation length $\xi_{AFM} \gg a$ at $x \ll 1$. Due to the small value of the electron mean-free path $l \sim a$, the main effect of the magnetic short range correlation is the same as that of the long range order: the electronic structure of the undoped CuO_2 layer is formed mainly by the hopping inside the antiferromagnetic cluster (Dagotto et al 1995). The dynamical character of the short range order results in some deviations from the mean field picture, i.e. instead of the Unklampf processes in the reduced antiferromagnetic Brillouin zone with the equivalent k and $k+Q$ ($Q = (\pi/a, \pi/a)$) vectors in the short order phase there are dynamical transitions between k and $k+Q$ states with the damping of the final $k+Q$ states. Thus, the antiferromagnetic Brillouin zone symmetry in Fig. 8.5 for $x = 0.10$ changes due to the shadow bands (Haas et al 1995). Our calculations for the undoped region of concentration $x_{cr} < x < x_{opt}$ may be considered as the interpolation between the long range antiferromagnetic state at $x < x_{cr}$ and the paramagnetic state at $x \sim x_{opt}$ where x_{opt} is an optimal doping concentration with $\xi_{AFM} \sim a$. The analysis of the two-magnon spectra dependence on the doping concentration in the Bi-2212 cuprates carried out by Guptasarma et al (1999) has shown that there is a crossover between the regimes of the doped antiferromagnet and the near antiferromagnet Fermi-liquid state (NAFL) at $x \approx x_{opt}$. A similar conclusion is given by the non-Fermi liquid effects due to the dynamical short range order in the underdoped region, as verified by the FLEX treatment of the Hubbard model (Schmalian et al 1996).

In the paramagnetic phase at $x \sim x_{opt}$, the band structure ($x = 0.2$ curve in Fig. 8.5) has a typical single electron spectrum with the maximum at the $M = (\pi/a, \pi/a)$ point and the van Hove singularity at the $X = (\pi/a, 0)$ point. A similar spectrum has been obtained using the quantum Monte-Carlo (QMC) calculations (Daffy et al 1997) and has been measured experimentally by the ARPES (Marshal et al 1996) on the optimally doped Bi-2212 samples.

The comparison of the valence band dispersion for the undoped antiferromagnetic insulator (Fig. 8.4) and underdoped metal (Fig. 8.5) shows that besides the shift of the maximum from the $(\pi/a, \pi/a)$ point there is one more important feature in the underdoped case, viz. a gap between the bands at $X = (\pi/a, 0)$ point. This gap gives rise to the pseudogap in the density of states as measured by the nuclear magnetic resonance, inelastic neutron scattering and ARPES methods (see e.g. review by Ovchinnikov (1997)). Our results on the concentration evolution of the valence band spectrum can be compared with the QMC calculations by Daffy (1997), and the works of Prelovsek et al (1999) on the single electron spectral function and the ARPES data. In our approach, the spectral density has a low energy peak with the dispersion given by Fig. 8.4 and high-energy satellite resulting from the impurity-like band. The low energy peak weakly depends on doping similar to the QMC result (Prelovsek et al 1999) in the t–J model. With decreasing temperature, the splitting of the quasiparticle peak in the spectral density in the low and high energy satellites has been found by Prelovsek et al (1999) at small doping concentration. A similar splitting has been obtained previously by Kampf and Schrieffer (1990) in the spin bag model. According to the NAFL model (Schmalian et al 1999), near the symmetrical points $X = (\pi/a, 0)$ and $Y = (0, \pi/a)$ there are hot electrons at the Fermi surface that are most sensitive to the spin correlations. In our calculations the comparison of the $x = 0.10$ curves for the antiferromagnetic and the paramagnetic phases also reveals the main difference at the X point with a pseudogap in the antiferromagnetic state.

According to the ARPES data (Wells *et al* 1995; Marshall *et al* 1996) the dispersion of the valence band in the direction $(0,0) \leftrightarrow (\pi/2a, \pi/2a)$ weakly depends on the concentration while there appear new states below the Fermi level with doping near the X point. Decreasing the concentration of holes from the optimal to the undoped region results in the gap formation along the XM line, directly proving the non-rigid behavior of the valence band with doping and the pseudogap at the X point.

We calculate the doping dependence of the chemical potential by solving the standard equation for the chemical potential

$$\frac{1}{N} \sum_{f\lambda\sigma} \langle a^+_{f\lambda\sigma} a_{f\lambda\sigma} \rangle = n_{hole} = 1 + x.$$

For $x \leq x_{opt} = 0.18$ we have considered the two-sublattice magnetic ordered state and for $x > x_{opt}$ we have considered the paramagnetic phase. Our results are given by Fig. 8.6 with the experimental dependence $\mu(x)$ obtained by precise measurement of the core-level photoemission spectra of $La_{2-x}Sr_xCuO_4$ (Harima *et al* 2001).

We found in agreement with the experiment that the chemical potential is pinned, up to the optimal doping at the top of the valence band. The reason for the pinning is the unusual properties of the in-gap state with the spectral weight proportional to x. This narrow impurity-like band contains x states (contrary to 2 states for free electron band) and is filled with x

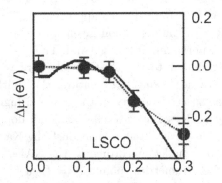

Fig. 8.6 The concentration dependence of the chemical potential in $La_{2-x}Sr_xCuO_4$ calculated by GTB method (Borisov *et al* 2002) and measured by ARPES experiment (Harima *et al* 2001).

extra holes. This quasiparticle disappears in paramagnetic phase, which is why $\mu(x)$ decreases at $x > x_{opt}$.

We have calculated also the Fermi surface at different doping levels. For the underdoped region with the extrapolation of the antiferromagnetic phase we found no agreement between calculation and the experiment. The shape of the Fermi surface is more subtle than the chemical potential which is the integral characteristic of the band structure. The second remark is that the optimal doping concentration $x_{opt}^{exp} = 0.18$ is determined by the van Hove singularity. Our calculation for the paramagnetic phase gives a concentration of holes where μ coincides with the calculated van Hove energy $x_{opt}^T \approx 0.55$. We assume that the doping dependence itself is correct and only the scale is wrong due to the omission of spin fluctuations in the paramagnetic phase. Thus we introduce the relative concentration of holes

$$x^T = \frac{x^* x_{opt}^T}{x_{opt}^{exp}}.$$

The Fermi surface calculated for the relative concentrations are given for the overdoped regime in Fig. 8.7 (Borisov et al 2002).

The good agreement to the ARPES data for $La_{2-x}Sr_xCuO_4$ (Ino et al 2002) proves that our scaling procedure is reasonable. The lower the concentration the larger the disagreement with ARPES data. We explain it by resorting to the necessity to invoke a more accurate description of the spin system in the underdoped region. Particularly, contribution of the short magnetic order (spin–liquid correlations) in the paramagnetic phase results

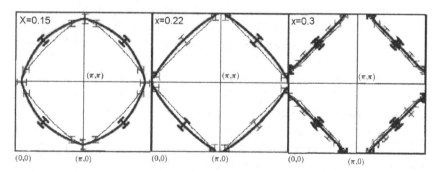

Fig. 8.7 The doping dependence of the Fermi surface in $La_{2-x}Sr_xCuO_4$ calculated by GTB method (Borisov et al 2002) and measured by ARPES experiment (Ino et al 2002).

in a shift of the van Hove singularity and, concequently, in the decrease of x_{opt}^T.

Concluding the discussion on the electronic structure of copper oxides we can state that the generalized tight-binding approach in the framework of the multiband p–d model provides a rather realistic description of the quasiparticle spectra both in the undoped and doped cases. The doping dependence of the valence band is quite unusual as compared to the rigid model type behavior in the $(0,0) \leftrightarrow (\pi/2a, \pi/2a)$ direction and the non-rigid behavior near the $X = (\pi/a, 0)$ point of the Brillouin zone. Qualitatively, the dispersion is similar to the t–J model but, quantitatively, there is a difference near the X point resulting from the two-hole triplet excitations. We have shown that the smooth evolution of the electronic band structure from the charge-transfer insulator undoped state to the paramagnetic overdoped metal state occurs via the formation of an unusual in-gap state and its evolution forward the narrow impurity like band. The chemical potential is a rather robust integral characteristic and its value is not sensitive to the details of the spin correlations. The Fermi surface is a more subtle characteristic and is in the agreement to ARPES data only in overdoped region.

8.4 Comparison of the Electronic Structure and Superconductivity of Cuprates and Ruthenates

More than a decade of intensive research of the cuprate superconductors and related systems has raised fundamental challenges to our understanding of the mechanism of high-temperature superconductivity. One of the most important question is, what is so specific in copper oxides, is it the unique chemistry of the planar Cu-O bond that determines the high value of T_c? The discovery of superconductivity in Sr_2RuO_4 with $T_c \sim 1$ K (Maeno et al 1994) is of particular interest because it has a similar crystal structure to the parent compound La_2CuO_4, one of the best studied families of the cuprate superconductors $La_{2-x}Sr_xCuO_4$, but has four valence electrons (for Ru^{4+}) instead of one hole per formula unit. It is generally believed that comparison of normal and SC properties of the cuprates and the ruthenate will give a deeper understanding of the nature of high-T_c superconductivity. While the normal state of doped cuprates looks almost like antiferromagnetic Fermi-liquid (Pines 1990), the normal state of Sr_2RuO_4 is characterized by the strong ferromagnetic fluctuations (Oguchi 1995). The properties of superconducting state are also different: the singlet pairing with major contribution of the $d_{x^2-y^2}$ symmetry was suggested for the cuprates (Bickers et al 1987), while the triplet pairing with p-type symmetry similar to the $^3HeA_1$ phase is proposed for Sr$_2$RuO$_4$ (Rice and Sigrist 1995). The triplet superconductivity in Sr$_2$RuO$_4$ is induced by the ferromagnetic spin fluctuations (Mazin and Singh 1997).

To compare the SC in Sr$_2$RuO$_4$ and cuprates, Kuz'min et al (2000) have proposed a t–J–I model containing both an indirect antiferromagnetic coupling J and a direct ferromagnetic coupling I between neighboring cations. This model is based on the electronic structure calculations. An important difference from the cuprates is that relevant orbitals to states near the Fermi energy are the Ru $d \in (d_{xy}, d_{yz}, d_{xz})$ and the O p π, instead of the Cu $d_{x^2-y^2}$ and the O p σ states. Due to σ-bonding in the cuprates, a strong p–d hybridization takes place resulting in the strong antiferromagnetic coupling J, while a direct $d_{x^2-y^2}$ Cu–Cu overlapping is negligible. In Sr$_2$RuO$_4$ with π-bonding, the Ru–O–Ru 180° antiferromagnetic superexchange coupling is weak (Goodenough 1963) while a direct d_{xy} Ru–Ru overlapping is not small. That is why the Heisenberg type direct Ru–Ru exchange interaction is added to the Hamiltonian of the t–J model. The strong electron correlations are common features of the charge carriers in cuprates and Sr$_2$RuO$_4$ in our model. These correlations for the cuprates are well known. The importance of electron correlations for Sr$_2$RuO$_4$ follows from the high value of the effective mass of electrons in the γ-band $m^* \sim 10 m_e$ obtained by the quantum oscillations measurements (Mackenzie et al 1996).

The Hamiltonian of the proposed t–J–I model is written in the form

$$H = H_{kin} + H_{int},$$

$$H_{kin} = \sum_{f\sigma}(\varepsilon - \mu)X_f^{\sigma\sigma} - t\sum_{f\delta\sigma}X_f^{\sigma 0}X_{f+\delta}^{0\sigma}, \quad (8.25)$$

$$H_{int} = J\sum_{f\delta}K_{f,f+\delta}^{(-)} - I\sum_{f\delta}K_{f,f+\delta}^{(+)},$$

$$K_{fg}^{(\pm)} = \boldsymbol{S}_f \cdot \boldsymbol{S}_g \pm \frac{1}{4}n_f n_g, \quad n_f = \sum_{f\sigma}X_f^{\sigma\sigma}.$$

This Hamiltonian is given on d-dimensional lattice of N sites (f is sites of the lattice), with z nearest neighbors (NN) and periodic boundary conditions (δ is vector connecting NN). It describes a system of N_e electrons in subspace of local states $|0\rangle$ and $|\sigma\rangle$, which are holes (empty sites of the lattice) and one-electron states with $\sigma = \uparrow$ and $\sigma = \downarrow$, so that $0 \le N_e \le N$. H_{kin} describes hopping of electrons onto nearest empty sites of the lattice (hopping integral $t > 0$), ε is energy of one-electron level[1], μ is chemical potential.

The antiferromagnetic ($J > 0$) and the ferromagnetic ($I > 0$) exchange interactions appear in the Hamiltonian $H_{int} \cdot S_f$ and n_f are, repectively, the spin operator with $S = 1/2$ and electron number operator on site f. The parameters J and I correspond to numerically value of the system's exchange energy per bond (between NN). For example, the energy of the saturated ferromagnetic state $|F\rangle$ with $n = 1$ is equal to

$$E_F = \left\langle F \left| H_{ex}^{(+)} \right| F \right\rangle = -I \cdot \frac{1}{2} zN,$$

and the energy of the antiferromagnetic state, calculated with the Neel wavefunction $|AF\rangle$, is equal to

$$E_{AF} = \left\langle AF \left| H_{ex}^{(-)} \right| AF \right\rangle = -J \cdot \frac{1}{2} zN,$$

where $zN/2$ is the full number of bonds between all NN. Thus, the parameters I and J already contain square of the spin $S = 1/2$. In general case, the exchange interaction between NN is the sum of the antiferromagnetic kinetic exchange $J_{kin} = 2t^2/U$ and the direct exchange which may be either antiferromagnetic or ferromagnetic. If both the exchanges are antiferromagnetic, then we assume $I = 0$ and use common parameter J. But if direct exchange is ferromagnetic, then J–I competition appears and this case is reflected in exchange part of the Hamiltonian H_{int}. For cuprates $J \gg I$ (or $I = 0$), and for Sr_2RuO_4 $I \gg J$.

To get superconductivity the copper oxides should be doped but Sr_2RuO_4 is self-doped. According to the band structure calculations (Singh 1995), the electron α-band in Sr_2RuO_4 is half-filled, the hole β-band has $n_0 = 0.28$ holes and the electron γ-band with d_{xy} contribution is more then half-filled, $n_\gamma = 1 + n_0$. In our model, the strong electron correlations split the γ-band into filled lower Hubbard band (LHB) with $n_e = 1$ and partially filled upper Hubbard band (UHB) with the electron concentration $n_e = n_0$. We use the hole representation where the electron UHB transforms into the hole LHB with hole concentration $n_h = 1 - n_0$. All other bands (α and β) are treated here as an electron reservoir. Observation of a square

---

[1] In the following, this energy is assumed to be 0.

flux-line lattice in Sr_2RuO_4 suggests that superconductivity resides mainly on the γ-band (Riseman et al 1998). For the cuprates the quasiparticle is a hole in the electron LHB with the electron concentration $n_e = 1 - n_0$; for $La_{2-x}Sr_xCuO_4$ $n_0 = x$.

For convenience, we normalize the Hamiltonian (8.25) with the free electron half bandwidth $W = zt$. After Fourier transforming Hubbard operators

$$X_{k\sigma} = \frac{1}{\sqrt{N}} \sum_f e^{ikf} X_f^{0\sigma}, \quad X_q^{\sigma\sigma'} = \frac{1}{\sqrt{N}} \sum_f e^{ikf} X_f^{\sigma\sigma'},$$

where the vectors k, q belong to the first Brillouin zone, we obtain the dimensionless Hamiltonian of our model in the form

$$h = H/zt = h_{kin} + h_{int},$$

$$h_{kin} = \sum_{k\sigma}(\omega_k - \tilde{\mu})X_{k\sigma}^+ X_{k\sigma}, \quad \omega_k = -\frac{1}{z}\sum_\delta e^{ik\delta} = -\gamma_k,$$

$$h_{int} = \frac{1}{2}\sum_q \gamma_q \left\{ g \sum_\sigma \left(X_q^{\sigma\bar\sigma} X_{-q}^{\bar\sigma\sigma} - X_q^{\sigma\sigma} X_{-q}^{\bar\sigma\bar\sigma}\right) \right.$$

$$\left. -r \sum_\sigma \left(X_q^{\sigma\bar\sigma} X_{-q}^{\bar\sigma\sigma} + X_q^{\sigma\sigma} X_{-q}^{\bar\sigma\bar\sigma}\right) \right\},$$

where $\sigma = -\bar\sigma$ and the dimensionless parameters $g = J/t, r = I/t$, and $\tilde{\mu} = \mu/zt$ have been used.

There are many ways to get the mean field solutions for the superconducting state, we have used the irreducible Green's function method (Tyablikov 1975; Plakida et al 1989) by projecting the higher-order Green's functions onto subspace of normal and abnormal Green's functions coupled via the Gor'kov system of equations.

Using the dimensionless Hamiltonian $h = H/zt$ and the algebra of the X-operators, we derive the equation of motion for the quasi-Fermi operator ($\hbar = 1$):

$$i\dot{X}_{k\sigma} = [X_{k\sigma}, h] = (\omega_k - \tilde{\mu})X_{k\sigma} + L_{k\sigma}, \quad (8.26)$$

$$L_{k\sigma} = L_{k\sigma}^{(kin)} + L_{k\sigma}^{(int)},$$

$$L_{k\sigma}^{(kin)} = \frac{1}{\sqrt{N}} \sum_p \omega_p \left(X_{k-p}^{-\sigma\sigma} X_{p\bar\sigma} - X_{k-p}^{\bar\sigma\bar\sigma} X_{p\sigma}\right),$$

$$L_{k\sigma}^{(int)} = \frac{1}{\sqrt{N}} \sum_p \gamma_{k-p} \left\{ (g-\lambda)X_{k-p}^{\bar\sigma\sigma} X_{p\bar\sigma} - g X_{k-p}^{\bar\sigma\bar\sigma} X_{p\sigma} - r X_{k-p}^{\sigma\sigma} X_{p\sigma} \right\},$$

where the nonlinear operator $L_{k\sigma}$ describes the electron correlations with both the opposite and the same spin projections.

Let us introduce the irreducible operator

$$\bar{L}_{k\sigma} = L_{k\sigma} - \frac{\langle\{L_{k\sigma}, X_{k\sigma}^+\}\rangle}{\langle\{L_{k\sigma}, X_{k\sigma}^+\}\rangle} X_{k\sigma} - \frac{\langle\{L_{k\sigma}, X_{-k\bar{\sigma}}\}\rangle}{\langle\{X_{-k\bar{\sigma}}^+, X_{-k\bar{\sigma}}\}\rangle} X_{-k\bar{\sigma}}^+$$

possessing the property of "orthogonality at the average":

$$\langle\{\bar{L}_{k\sigma}, X_{k\sigma}^+\}\rangle = \langle\{\bar{L}_{k\sigma}, X_{-k\bar{\sigma}}\}\rangle = 0.$$

Then, Eq. (8.26) can be written in the form:

$$i\dot{X}_{k\sigma} = \left(\omega_k - \tilde{\mu} + \frac{C_{k\sigma}}{1-n_{\bar{\sigma}}}\right)X_{k\sigma} + \frac{\Delta_{k\sigma}}{1-n_{\bar{\sigma}}}X_{-k\sigma}^+ + \bar{L}_{k\sigma}, \qquad (8.27)$$

where $C_{k\sigma} = \langle\{L_{k\sigma}, X_{k\sigma}^+\}\rangle$ and $\Delta_{k\sigma} = \langle\{L_{k\sigma}, X_{-k\bar{\sigma}}\}\rangle$. Here $C_{k\sigma}/(1-n_{\bar{\sigma}})$ determines the renormalization spectrum and $\Delta_{k\sigma}$ describes possible superconducting gap.

The generalized Hartree–Fock approximation (GHFA) or mean field approximation corresponds to the linear part of Eq. (8.27), i.e. without the irreducible operator $\bar{L}_{k\sigma}$. We shall consider the possible superconducting solutions neglecting the irreducible operator. The spectral renormalization can be calculated in the simplest approximation of the Hubbard 1 type. Then in the nonmagnetic ground state ($n_\uparrow = n_\downarrow = n/2$), the dependence on spin projection disappears and the modified spectrum can be represented in the form

$$\xi_k = c(n)(\omega_k - m),$$
$$m = [(g+r)n/4 + \tilde{\mu}]/c(n),$$
$$c(n) = 1 - n/2$$

where m is the effective chemical potential.

The expression for the gap $\Delta_{k\sigma}$ is reduced to the form

$$\Delta_{-k\downarrow} = -\Delta_{k\uparrow} = \Delta_k,$$
$$\Delta_k = \frac{1}{N}\sum_p [2\omega_p + g(\gamma_{k+p} + \gamma_{k-p}) - r\gamma_{k+p}]B_p, \qquad (8.28)$$

where $B_p = \langle X_{-p\downarrow}X_{p\uparrow}\rangle$ are the abnormal averages. The first term in Eq. (8.28) is caused by kinematic correlations of electrons and is derived from the kinetic term in the Hamiltonian (so-called kinematic mechanism of

The Electronic Structure of Copper Oxides in the Multiorbital p-d Model 223

pairing (Zaitsev and Ivanov 1987)). The rest is the effect of exchange interaction. In the mean field approximation, we derive the system of equations

$$i\dot{X}_{k\uparrow} = \xi_k X_{k\uparrow} - \frac{\Delta_k}{c(n)} X^+_{-k\downarrow},$$

$$i\dot{X}^+_{-k\downarrow} = -\xi_k X^+_{-k\downarrow} - \frac{\overset{*}{\Delta}_k}{c(n)} X_{k\uparrow}.$$

(8.29)

Using Eq. (8.29), we get the Gor'kov type system of equations for the retarding anticommutator Green's functions, and its solution:

$$\langle\langle X_{k\uparrow} | X^+_{k\uparrow} \rangle\rangle_E = c(n) \frac{E + \xi_k}{E^2 - E_k^2},$$

(8.30)

$$\langle\langle X^+_{-k\downarrow} | X^+_{k\uparrow} \rangle\rangle_E = \frac{\overset{*}{\Delta}_k}{E^2 - E_k^2},$$

where

$$E_k^2 = \xi_k^2 + \frac{|\Delta_k|^2}{c(n)^2}.$$

According to spectral theorem, we find the normal and the abnormal averages:

$$n_k = \langle X^+_{k\uparrow} X_{k\uparrow} \rangle = \langle X^+_{k\downarrow} X_{k\downarrow} \rangle = c(n) \cdot \frac{1}{2} \left(1 - \frac{\xi_k}{E_k} \tanh \frac{E_k}{2\tau} \right) \equiv c(n) \cdot f_k,$$

(8.31)

$$B_k^* = \langle X^+_{k\uparrow} X^+_{-k\downarrow} \rangle = \frac{\overset{*}{\Delta}_k}{2E_k} \tanh \frac{E_k}{2\tau},$$

(8.32)

where $E_k > 0$, and $\tau = k_B T/zt$ is the dimensionless temperature.

In the superconducting phase we have a system of three self-consistency equations:

(i) The interdependence of concentration n and effective chemical potential m based on Eq. (8.32) for the normal averages.
(ii) The equation for the energy gap Δ_k which is defined by Eq. (8.28) and obeys condition (8.31).
(iii) A sum rule for the abnormal averages B_p as a result of the algebra of X-operators in the subspace of states $|0\rangle, |\sigma\rangle$.

The multiplication rule for X-operator results in

$$\frac{1}{N}\sum_k B_k = \frac{1}{N}\sum_k \langle X_{-p\downarrow} X_{p\uparrow}\rangle = \frac{1}{N}\sum_f \langle X_f^{0\downarrow} X_f^{0\uparrow}\rangle = 0. \tag{8.33}$$

This constraint is the direct result of the strong electron correlation and is the most essential difference between our approach and the traditional mean field theory.

We will demonstrate just which solutions exist and which solutions obey all the three equations.

Let us represent the abnormal averages $B_p \equiv \langle X_{-p\downarrow} X_{p\uparrow}\rangle$ in the form:

$$B_k = B_k^{(s)} + B_k^{(a)},$$

$$B_k^{(s)} = \frac{1}{2}(B_k + B_{-k}) = B_{-k}^{(s)},$$

$$B_k^{(a)} = \frac{1}{2}(B_k - B_{-k}) = -B_{-k}^{(a)},$$

i.e. as a sum of symmetric (s) and antisymmetric (a) parts. At once we remark that the sum rule (8.33) is fulfilled automatically for the antisymmetric part $B_k^{(a)}$. It is easily to show that $B_k^{(s)}$ describes the singlet pairing and $B_k^{(a)}$ describes the triplet pairing with $S^z=0$.

We consider the alternant lattices ($z = 2d, d = 2, 3$) for which

$$\gamma_k = \frac{1}{d}\sum_j \cos k_j$$

(lattice parameter $a = 1$). Since

$$\gamma_{k\pm p} = \frac{1}{d}\sum_j (\cos k_j \cos p_j \pm \sin k_j \sin p_j),$$

the gap can be represented in the form

$$\Delta_k = \Delta_k^{(s)} + \Delta_k^{(a)},$$

$$\Delta_k^{(s)} = 2\Delta_0 + (2g - r)\frac{1}{d}\sum_j C_j \cos k_j,$$

$$\Delta_0 = \frac{1}{N}\sum_p \omega_p B_p^{(s)},$$

$$\Delta_k^{(a)} = r\frac{1}{d}\sum_j S_j \sin k_j,$$

where

$$C_j = \frac{1}{N}\sum_p \cos p_j B_p^{(s)}, \quad S_j = \frac{1}{N}\sum_p \sin p_j B_p^{(a)}.$$

Here the gap $\Delta_k^{(s)}$ corresponds to the singlet pairing and $\Delta_k^{(a)}$ to the triplet pairing.

In principle, several solutions can exist, and each them is a definite combination of cosines (for S-pairings) and sines (for T-pairings). In the general case, we number the solutions with symbol l and denote the gap of l-type as Δ_{kl} and the spectrum as E_{kl}.

(1) Symmetric solution for S-pairing of s-type, $l = 0$. If $C_x = C_y = C_z$ for $d = 3$ or $C_x = C_y$ for $d = 2$, then we have

$$\Delta_{k0} = (2 + \lambda_0 \omega_k)\Delta_0, \quad \lambda_0 = 2g - r, \quad (8.34)$$

where λ_0 is a dimensionless coupling constant for singlet (S) pairing. In this case, the sum of the abnormal averages (while $T = 0$) is

$$\frac{1}{N}\sum_p B_{p0} = \Delta_0 = \frac{1}{N}\sum_p \frac{2 + \lambda_0 \omega_p}{\sqrt{\xi_p^2 + |\Delta_{p0}|^2/c^2(n)}} \neq 0.$$

The constraint condition (8.33) is not fulfilled for the abnormal averages B_{k0} with gap Δ_{k0} and for this reason the solutions of s-type are not present (Plakida et al 1989, Izyumov Yu and Letfulov 1991).

Furthermore, we restrict ourselves to the analysis of the $d = 2$ case. The antisymmetric solution of p-type ($l = 1$) and the symmetric solution of d-type ($l = 2$) can be represented in a single form

$$\Delta_{kl} = \lambda_l \psi_l(k)\Delta_l, \quad \Delta_l = \frac{1}{N}\sum_p \psi_l(p) B_{pl}. \quad (8.35)$$

We have the next types of solutions in the explicit form:

(2) Antisymmetric solution of p-type (triplet pairing), $l = 1$:

$$\psi_p(k) = \frac{1}{2}(\sin k_x + i \cdot \sin k_y), \quad \lambda_p = r. \tag{8.36}$$

(3) Symmetric solution of d-type (singlet pairing), $l = 2$:

$$\psi_d(k) = \frac{1}{2}(\cos k_x - \cos k_y), \quad \lambda_d = 2g - r. \tag{8.37}$$

The sum rule (8.33) can be written as

$$\frac{1}{N}\sum_p \frac{\psi_l(p)}{E_{pl}} = 0$$

and is fulfilled automatically for the p-type and can be proven easily for the d-type using the symmetry properties.

The gap equations for the p–d-states are

$$\frac{1}{\lambda_l} = \frac{1}{N}\sum_p \frac{|\psi_l(p)|^2}{2E_{pl}}\tanh\left(\frac{E_{pl}}{2\tau}\right), \tag{8.38}$$

where

$$E_{pl} = \sqrt{c^2(n)(\omega_p - m)^2 + \frac{|\Delta_{pl}|^2}{c^2(n)}}.$$

The equation for T_c in p- and d-states is given by

$$\frac{2c(n)}{\lambda_l} = \frac{1}{N}\sum_p \frac{|\psi_l(p)|^2}{|\omega_p - m|}\tanh\left(\frac{c(n)|\omega_p - m|}{2\tau_c^{(l)}}\right). \tag{8.39}$$

The same equation for the $d_{x^2-y^2}$ pairing has been derived by the diagram technique for the t–J model by Izyumov and Letfulov (1991).

At the numerical solution of the Eq. (8.39) more then 10^6 points of the Brillouin zone have been taken. Results of the computations $T_c(n)$ are shown in the Fig. 8.8 for several values of the coupling constants λ_l. These results have revealed the remarkable difference in T_c values: $T_c^{(p)} \ll T_c^{(d)}$ when $\lambda_p = \lambda_d$. The moderate values of $\lambda \approx 0.4$–0.5 and $zt \approx 0.5\,\text{eV}$ result in $T_c^{(p)} \sim 1\,K, T_c^{(d)} \sim 100\,K$. The mean field stability criterion for the p pairing is $\lambda_p > 0$ and for the d pairing is $\lambda_d > 0$. As for the stability of the mean field solution to the charge density and spin fluctuations, recently this stability have been proved using the same approach by Plakida and Oudovenko (1999) where the self-energy corrections was calculated.

The Electronic Structure of Copper Oxides in the Multiorbital p-d Model 227

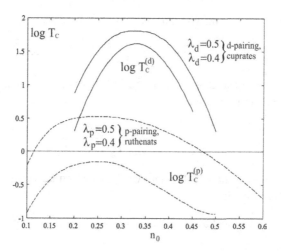

Fig. 8.8 Concentration dependence of T_c for the triplet pairing of the p-type and singlet pairing of the d-type, $\lambda_p = I/t, \lambda_p = (2J - I)/t, I(J)$ is a parameter of ferro (antiferro)magnetic exchange, t is the hopping integral. Here $t = 0.1\,\mathrm{eV}$.

It is clear from the definitions (8.36), (8.37) that the p-type superconductivity is formed by the ferromagnetic interaction in the case of $\mathrm{Sr_2RuO_4}$, and that the d-type superconductivity is induced by the antiferromagnetic interaction in copper oxides. The d-type cannot arise in ruthenates because of the inappropriate sign of the parameter λ_d, the p-type is displaced by the d-type in the cuprates for the reason that the λ_p is very low. To understand why $T_c^{(p)} \ll T_c^{(d)}$, we have analysed Eq. (8.30) analytically. Using integration over the constant energy surfaces $\omega_k = \omega$ it can be written as

$$\frac{2c(n)}{\alpha_l} = \int_{-1}^{+1} \frac{\psi_l^2(\omega)}{|\omega - m|} \tanh\left(\frac{c(n)\omega - m|}{2\tau_c}\right) d\omega,$$

$$\psi_l^2(\omega) = \frac{1}{(2\pi)^2} \oint_{(\sigma_\omega)} \frac{|\psi_l(k)|^2}{|\nabla_k \omega_k|} d\sigma_\omega. \tag{8.40}$$

The sum rule for the $\psi_l^2(\omega)$ functions is the same for $l = p$ and $l = d$

$$\frac{1}{N}\sum_k |\psi_l(k)|^2 = \int_{-1}^{+1} \psi_l^2(\omega) d\omega = 1/4.$$

For the p-state

$$\frac{|\psi_p(k)|}{|\nabla_k \omega_k|} = |\nabla_k \omega_k|$$

and $\psi_p^2(\omega)$ is rather small with smooth energy dependence,

$$\psi_p^2(\omega) = \frac{2}{\pi^2}\left[E\left(\sqrt{1-\omega^2}\right) - \omega^2 K\left(\sqrt{1-\omega^2}\right)\right] \approx \frac{2}{\pi^2}(1-|\omega|^{1.61}),$$

where K and E are the Legendre complete elliptic integrals of the first and the second kind. For the d-state,

$$\frac{|\psi_d^2(k)|}{|\nabla_k \omega_k|} = \frac{1}{2}\frac{(\cos k_x - \cos k_y)^2}{\sqrt{\sin^2 k_x + \sin^2 k_y}},$$

there is the same singularity like the van Hove singularity in the density of states $\rho(\omega)$. The result of calculation is

$$\psi_d^2(\omega) = (1-\omega^2)\rho(\omega) - 2\psi_p^2(\omega),$$

where

$$\rho(\omega) = \frac{2}{\pi^2}K\left(\sqrt{1-\omega^2}\right) \approx \frac{1}{\pi} - \left(\frac{1}{2} - \frac{1}{\pi}\right)\ln(|\omega|).$$

The comparison of $\psi_p^2(\omega)$ and $\psi_d^2(\omega)$ has shown that the van Hove singularity is cancelled in the p-state but is not cancelled in the d-state (Fig. 8.9).

A similar conclusion on the large contribution due to van Hove singularity in the case of $d_{x^2-y^2}$ pairing in the strongly correlated holes in the

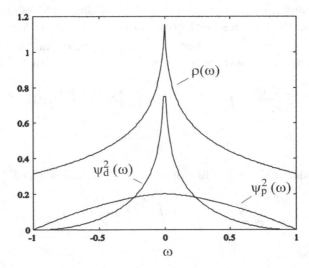

Fig. 8.9 Energy dependence of the effective gap anisotropy ψ_p^2 and ψ_d^2 and single-particle density of states $\rho(\omega)$ in square lattice.

CuO$_2$ plane have been obtained previously in the framework of the t–J model (Dagotto et al 1995) and the Hubbard model (Beenen and Edwards 1995).

The parameters t and I can be estimated from some experimental data. While the concentration of "double states" is low in the upper Hubbard subband (these states are equivalent to holes in the lower Hubbard subband), the dispersion law is well described in effective mass approximation:

$$\varepsilon_k = -zt\gamma_k \approx \varepsilon_0 + p^2/2m^*, \quad 1/m^* = 2ta^2/\hbar^2,$$

where a is distance between atoms. Substituting values of the effective mass $m^* = (5\text{--}10)m_e$ and distances $a = (2\text{--}4)$ angstrom, we obtain t in interval $t = (0.02\text{--}0.2)$ eV with typical value of $t\sim 0.1$ eV.

The spin-wave theory formula $k_B T_C \sim zI_{(xy)}/\ln[I_{(xy)}/I_{(z)}]$ may be used for the Curie temperature T_C. Here $I_{(xy)} \equiv I$ is the intraplane exchange and $I_{(z)}$ is the interplane exchange. The ratio of this interactions may be $10^4\text{--}10^5$ and the Curie temperature T_C may reach the value of $\sim(100\text{--}200)$ K while $I \sim (2\text{--}5) \times 10^{-2}$ eV. Thus, the ruthenates' values $r = I/t \sim (0.2\text{--}0.5)$ are equal roughly to the cuprates' dimensionless parameter $\lambda_d = 2J/t$.

In conclusion we have presented the model of strongly correlated electrons in two dimensional lattice that allows one to consider the cuprates ($J \gg I$) and Sr$_2$RuO$_4$ ($J \ll I$) on the same footing. The singlet SC in the s-state is absent in the strong correlation limit, the triplet p-pairing occurs due to the ferromagnetic fluctuations and the singlet d pairing is induced by the antiferromagnetic fluctuations. The reason why T_c is much higher in the cuprates than in Sr$_2$RuO$_4$ is due to the different gap anisotropy. For the p-state, the k-dependence of the gap ensure a cancellation of the van Hove singularity whereas for the d-state the gap anisotropy permits large van Hove singularity contribution in the equation for T_c. As for the question of high-T_c superconductivity in the copper oxide, the possible answer may be as follows: It is the planar Cu–O σ-bonding which results in the strong antiferromagnetic Cu–Cu interaction, and this interaction induces the singlet pairing with $d_{x^2-y^2}$ symmetry.

References

A. A. Abrikosov, L. P. Gor'kov and I. E. Dzyaloshinskii, *Quantum Field Theoretical Methods in Statistical Physics* (Pergamon Press, New York 1965).

A. Aharony, R. J. Birgeneau, A. Coniglio, M. A. Kastner and H. E. Stanle, *Phys. Rev. Lett.* **60**, 1330 (1988).

O. K. Andersen, A. Liechtenstein, O. Jepsen and F. Paulsen, *J. Phys. Chem. Solids* **56**(12), 1573 (1995).

P. W. Anderson, *Phys. Rev.* **124**, 41 (1961).

P. W. Anderson, *Phys. Rev. Lett.* **64**, 1839 (1990).

P. W. Anderson, *Science* **235**, 1196 (1987).

A. V. Andreev, K. P. Belov, A. V. Deryagin, Z. A. Kazei, R. Z. Levitin, A. Menovski, Yu. F. Popov and V. I. Cilantiev, *JETP* **75**, 2351 (1978).

V. I. Anisimov, J. Zaanen and O. K. Andersen, *Phys. Rev. B* **44**, 943 (1991).

P. V. Avramov and S. G. Ovchinnikov, *Zh. Struct. Khim.* **40**, 131 (1990).

A. D. Balaev, V. A. Gavrichkov, S. G. Ovchinnikov, V. K. Chernov, T. G. Aminov and G. G. Shabunina, *JETP* **86**(5), 1026 (1998).

A. Bansil and M. Lindroos, *J. Phys. Chem. Solids* **59**, 1879 (1998).

A. F. Barabanov, K. A. Kikoin and L. A. Maksimov, *Teor. Mat. Fiz.* **20**, 364 (1974).

A. F. Barabanov, K. A. Kikoin and L. A. Maksimov, *Teor. Mat. Fiz.* **25**, 87 (1975).

A. F. Barabanov, L. A. Maksimov and G. U. Uimin, *Zh. Eksper. Theor. Fiz.* **69**, 371 (1989).

A. F. Barabanov, R. O. Kuzian and L. A. Maksimov, *J. Phys.: Condens. Matter* **39**, 129 (1991).

B. Barbara, J. X. Boucherle, M. E. Rossignol and J. Schweizer, *Physica B* **86–88**, 83 (1977).

V. G. Bar'yakhtar, V. N. Krivoruchko and D. A. Yablonskii, "Green's Functions in Magnetism Theory" [in Russian], (Nauka, Dumka, Kiev, 1984).

J. Beenen and D. M. Edwards, *Phys. Rev. B* **52**, 13626 (1995).

V. I. Belinicher, A. L. Chernyshev and V. A. Shubin, *Phys. Rev. B* **53**, 335 (1996).

A. Bianconi, M. de Santis and A. M. Flank, *Physica C* **153–155**(1–4), 1760 (1988).
N. E. Bickers, D. J. Scalapino and S. R. White, *Phys. Rev. Lett.* **62**, 961 (1989).
A. S. Borovik-Romanov, *Antiferromagnetism, in Antiferromagnetizm I ferrity (Antiferromagnetism and Ferrites)*, Ya. G. Dorfman, Ed. (Akad. Nauk SSSr, Moscow, 1962).
W. Brenig, *Phys. Rep.* **251**, 153 (1995).
L. N. Bulayevsky, E. L. Nagaev and D. I. Khomsky, *Zh. Eksp. Teor. Fiz.* **54**, 1562 (1968) [*Sov. Phys. JETP*, **27**, 836 (1968)].
K. H. J. Buschow, *Rep. Progr. Phys.* **42**, 1373 (1979).
C. Castellani, C. Di Castro, D. Feinberg and J. Ranniger, *Phys. Rev. Lett.* **43**, 1957 (1979).
K. A. Chao, J. Spalek and A. M. Oles, *J. Phys. C* **10**, L271 (1977).
B. Coqblin and J. R. Schrieffer, *Phys. Rev.* **185**, 847 (1969).
R. Chitra and G. Kotliar, *Phys. Rev. B* **62**, 12715 (2000).
D. Daffy, A. Nazarenko, S. Haas, A. Moreo, J. Riera and E. Dagotto, *Phys. Rev.* **56**, 5597 (1997).
E. Dagotto, F. Ortolani and D. Scalapino, *Phys. Rev. B* **46**, 3183 (1992).
E. Dagotto, *Rev. Mod. Phys.* **66**, 763 (1994).
E. Dagotto, A. Nazarenko and A. Moreo, *Phys. Rev. Lett.* **74**, 310 (1995).
A. Damasccelli, D. H. Lu and Z.-X. Shen, *J. Electron Spectrosc. Relat. Phenom.* **117–118**, 165 (2001).
V. V. DruJinin and Yu. P. Irkhin, *Zh. Eksp. Teor. Fiz.* **51**, 1856 (1966).
E. J. Dyson, *Phys. Rev.* **102**, 1230 (1956).
I. E. Dzyaloshinskii, *Zh. Eksp. Teor. Fiz.* **47**, 336 (1964) [*Sov. Phys. JETP* **20**, 223 (1965)].
R. Eder and Y. Ohta, *Phys. Rev. B* **50**, 10043 (1994).
R. Eder, J. von der Brink and G. A. Sawatzky, *Phys. Rev. B* **54**, 732 (1996).
V. J. Emeri, *Phys. Rev. Lett.* **58**, 2794 (1987).
V. J. Emeri and G. Reiter, *Phys. Rev. B* **38**, 11938 (1998).
M. Sh. Erukhimov and S. G. Ovchinnikov, *Phys. Stat. Sol. B* **123**, 105 (1984).
H. Eskes, L. J. Tjeng and G. A. Sawatzky, *Phys. Rev. B* **41**, 288 (1990).
H. Eskes and G. A. Sawatzky, *Phys. Rev. B* **44**, 9656 (1991).
P. Fazekas, *Lecture Notes on Electron Correlation and Magnetism* (World Scientific, Singapore, 1999).
L. F. Feiner, J. H. Jefferson and R. Raimondi, *Phys. Rev. B* **53**, 8751 (1993).
M. E. Foglio and L. M. Falikov, *Phys. Rev. B* **20**, 4554 (1979).
P. Fulde, *Electron Correlations in Molecules and Solids* (Springer-Verlag, Berlin-Heidelberg, 1991).
P. Fulde and M. Loewenhaupt, *Adv. Phys.* **34**, 589 (1986).
Ya. B. Gaididei and V. M. Loktev, *Phys. Stat. Sol. B* **147**, 307 (1988).
D. A. Garanin and V. S. Lutovinov, *Fiz. Tverd. Tela. (Leningrad)* **26**, 2821 (1984) [*Sov. Phys. Sold State* **26**, 1706 (1984)].
M. Gaudin, *Nucl. Phys.* **15**, 89 (1960).
V. A. Gavrichkov, M. SH. Yerukhimov, S. G. Ovchinnikov and I. S. Edelman, *Zh. Eksp. Teor. Fiz.* **90**(4), 1275 (1986) [*Sov. Phys. JETP*, 1986].

V. A. Gavrichkov, S. G. Ovchinnikov, A. A. Borisov and E. G. Goryachev, Zh. Eksp. Teor. Fiz. **118**, 422 (2000) [*JETP* **91**, 369 (2000)].
V. A. Gavrichkov, S. G. Ovchinnikov and A. A. Borisov, *Phys. Rev.* B **64**(23), 235124 (2001).
F. Gebhard, *The Mott-Hubbard Metal Insulator Transition: Models and Methods* (Springer, Berlin, 1997).
A. Georges, G. Kotliar, W. Krauth and M. Rozenberg, *Rev. Mod. Phys.* **68**, 13 (1996).
J. B. Goodenough, *Magnetism and Chemical Bond* (Wiley, New York, London, 1963).
L. P. Gor'kov, Zh. Eksp. Teor. Fiz. **34**, 735 (1958) [*Sov. Phys. JETP.* **17**, 505 (1958)].
Y. Guo, J. M. Langlois and W. A. Goddard, *Science* **239**, 896 (1988).
A. G. Gurevich, *Magnitnyi rezonans v ferritakh I antiferromagnetikakh (Magnetic Resonance in Ferrites and Antiferromagnets)*, (Moscow, Nauka, 1973).
M. Guptasarma, D. G. Hinks and M. V. Kleim, *Phys. Rev. Lett.* **82**, 5349 (1999).
M. C. Gutzwiller, *Phys. Rev. Lett.* **10**, 159 (1963).
M. C. Gutzwiller, *Phys. Rev.* **134**, A923 (1964).
M. C. Gutzwiller, *Phys. Rev.* **137**, A1726 (1965).
S. Haas, A. Moreo and E. Dagotto, *Phys. Rev. Lett.* **74**, 4281 (1995).
F. D. M. Haldane, *J. Phys. C* **14**, 2585 (1981).
A. B. Harris and R. V. Lange, *Phys. Rev.* **157**, 296 (1967).
R. Hayn, A. F. Barabanov, J. Schulenburg and J. Richter, *Phys. Rev.* B **53**, 11714 (1996).
L. Hedin and S. Ludquist, *Solid State Physics* **23**, 1 (Academic, New York, 1969).
J. E. Hirsch, *Phys. Rev. Lett.* **54**, 1317 (1985).
J. E. Hirsch, *Phys. Lett.* A **136**, 163 (1989).
P. Hohenberg and W. Kohn, *Phys. Rev.* **136**, 864 (1964).
P. Horsch and W. Stephan, in *Electronic Properties of High-T_c Superconductors* (Kirchberg, Tirol, 1992), Eds H. Kuzmany, M. Mekring, J. Fink (Springer Series in Solid State Science, 1993). v. 113, p. 351.
J. C. Hubbard, *Proc. Roy. Soc.* A **276**, 238 (1963).
J. C. Hubbard, *Proc. Roy. Soc.* A **281**, 401 (1964).
J. C. Hubbard, *Proc. Roy. Soc.* A **285**, 542 (1965).
M. Hunt, P. Meeson, P. A. Probst, P. Reinders, M. Springford, W. Assmus and W. Sun, *J. Phys.: Condens. Matter* **2**, 6859 (1990).
S. Hufner, *Photoelectron Spectroscopy* (Springer, Berlin, 1995).
M.R. Ibarra, E. W. Lee, A. del Molal and O. Moze, *Solid State Comm.* **50**, 147 (1985).
M. Imada, A. Fujimori and Y. Tokura, *Rev. Mod. Phys.* **70**, 1039 (1998).
Yu. P. Irkhin, Zh. Eksp. Teor. Fiz. **50**, 379 (1966).
Yu. P. Irkhin, V. V. Drujinin and A. A. Kazakov, Zh. Eksp. Teor. Fiz. **54**, 1183 (1968).
Yu. P. Irkhin, *Usp. Fiz. Nauk.* **154**, 321 (1988).
Yu. A. Izyumov and F. A. Kassan-ogly, *Fiz. Met. Metalloved.* **30**, 229 (1970a).
Yu. A. Izyumov and F. A. Kassan-ogly, *Fiz. Met. Metalloved.* **30**, 449 (1970b).

Yu. A. Izyumov and M. V. Medvedev, *Zh. Eksp. Teor. Fiz.* **59**, 553 (1970).
Yu. A. Izyumov, F. A. Kassan-ogly and Yu. N. Skryabin, *Field Methods in the Theory of Ferromagnetism* (Nauka, Moscow 1974) [in Russian].
Yu. A. Izyumov and Yu. N. Skryabin, *Statisticheskaya mekhanika magnitouporyadochennykh System (Statistical Mechanics of Magnetically Ordered System)* (Nauka, Moscow, 1987).
Yu. A. Izyumov and B. M. Letfulov, *J. Phys.: Condens. Matter* **3**, 5373 (1991).
Yu. A. Izyumov, M. I. Katsnel'son and Yu. N. Skryabin, *Magnetism Kollektivisirovannykh Elektronov (Itinerant Electron Magnetism)* (Nauka, Moscow, 1994), p. 368.
Yu. A. Izyumov, *Usp. Fiz. Nauk.* **165**, 403 (1995) [*Phys. Usp.* **38**, 385 (1995)].
Yu. A. Izyumov, *Usp. Fiz. Nauk.* **167**, 465 (1997).
J. H. Jefferson, H. Eskes and L. F. Feiner, *Phys. Rev. B* **45** 7959 (1992).
L. P. Kadanoff and G. Baym, *Quantum Statistical Mechanics* (W. A. Benjamin, Inc. New York, 1962).
H. Kamimura and M. Eto, *J. Phys. Soc. Jap.* **59**, 3053 (1990).
A. Kampf and J. R. Schrieffer, *Phys. Rev. B* **41**, 6399 (1990).
A. P. Kampf, *Phys. Rep.* **249**, 219 (1994).
T. Kasuya, T. Suzuki and Y. Haga, *J. Phys. Soc. Jap.* **62**, 2549 (1993).
T. Kasuya, *J. Phys. Soc. Jap.* **64**, 1453 (1995).
M. I. Katsnelson and A. I. Lichtenstein "First-principles calculations of magnetic interactions in correlated systems". *Phys. Rev. B* **61**, 8906 (2000).
G. G. Khaliulin, *Pis'ma Zh. Eksp. Teor. Fiz.* **52**, 999 (1990) [*JETP Lett.* **52**, 389 (1990)].
G. G. Khaliulin and P. Horsch, *Phys. Rev. B* **47**, 463 (1993).
R. S. Knox and A. Gold, *Symmetry in the Solid State* (W.A. Benjamin, Inc. Amsterdam, 1964).
W. Kohn and L. J. Sham, *Phys. Rev.* **140**, 1133 (1965).
G. Kotlair and A. E. Ruckenstein, *Phys. Rev. Lett.* **57**, 1362 (1986).
V. G. Kukharenko, *Zh. Eksp. Teor. Fiz.* **69**, 632 (1975) [*Sov. Phys. JETP* **42**, 321 (1975)].
K. G. Kurtovoi and R. Z. Levitin, *Usp. Fiz. Nauk.* **153**, 193 (1987) [*Sov. Phys. Usp.* **30**, 827 (1987)].
E. V. Kuz'min, S. G. Ovchinnikov and I. O. Baklanov, *Phys. Rev. B* **61** (2000).
L. D. Landau and E. M. Lifshitz, *Statistical Physics*, 3rd edn. (Pergamon Press, Oxford, 1980).
G. H. Lander, A. Delapalme, P. J. Brown, J. C. Spirlet, J. Rebizant and O. Vogt, *J. Appl. Phys.* **57**, 3748 (1985).
A. I. Lichtenstein and M. I. Katsnelson, *Phys. Rev. B* **57**, 6884 (1998).
E. H. Lieb and F. Y. Wu, *Phys. Rev. Lett.* **20**, 1445 (1968).
I. M. Lifshits, M. Ya. Azbel and M. I. Kaganov, *Electron Theory of Metals* (Plenum Press, New York, 1973).
P. A. Lingaard, *Theory of Spin Excitations in Rare-Earth Systems* (Roskilde, Denmark, Riso National Laboratory, 1986).
Z. Liu and E. Manousakis, *Phys. Rev. B* **45**, 2425 (1992).
V. M. Loktev and V. S. Ostrovskiy, *Ukr. Fiz. Zh.* **23**, 1708 (1978) [in Russian].

S. V. Lovtsov and V. Yu Yushankhai, *Physica C.* **179**, 159 (1991).
U. Lundin, I. Sandalov, O. Eriksson and B. Johansson, *Solid State Commun.* **115**, 7 (2000).
V. S. Lutovinov and M. Yu. Reizer, *Zh. Eksp. Teor. Fiz.* **77**, 707 (1979) [*Sov. Phys. JETP* **50**, 355 (1979)].
J. M. Luttinger, *Math. Phys.* **4**, 1154 (1968).
A. P. Mackenzie, S. R. Julian, A. J. Diver, G. J. McMillan, M. P. Ray, G. G. Lonzarich, Y. Maeno, Nishizakis and T. Fujita, *Phys. Rev. Lett.* **76**, 3786 (1996).
Y. Maeno, H. Hasimoto, K. Yoshida, S. Nishizaki, T. Fujita and F. Lichtenberg, *Nature (London)* **372**, 532 (1994).
Th. A. Maier, Th. Pruschke and M. Jarrell, *Phys. Rev. B* **66**, 075102 (2002).
D. S. Marshall, D. S. Dessau, A. G. Loeser, C.-H. Park, A. Y. Matsura, J. H. Eckstein, I. Bozovic, P. Fournier, A. Kapitulnik, W. E. Spicer and Z.-X. Shen, *Phys. Rev. Lett.* **76**, 4841 (1996).
G. Martinez and P. Horsch, *Phys. Rev. B* **44**, 317 (1991).
I. I. Mazin and D. J. Singh, *Phys. Rev. Lett.* **79**, 733 (1997).
W. Metzner and D. Vollhardt, *Phys. Rev. Lett.* **62**, 324 (1989).
N.F. Mott, *Proc. Phys. Soc. A* **62**, 416 (1949).
N. F. Mott, *Metal-Insulator Transitions* (Taylor and Francis, London, 1974).
E. Muller-Hartmann, *Z. Phys. B* **74**, 507 (1989).
E. L. Nagaev, *Zh. Eksp. Teor. Fiz.* **58**, 1269 (1970).
E. L. Nagaev, *Fiz. Tverd. Tela. (Leningrad)* **13**, 958 (1971).
E. L. Nagaev, *Physics of Magnetic Semiconductors* (Mir, Moscow, 1983) [in Russian].
E. L. Nagaev, *Magnetiki so slozhnymi obmennymi vzaimodeistviyami (Magnetic Materials with Complex Exchange Interactions)*, (Nauka, Moscow, 1988).
Y. Nagaoka, *Phys. Rev.* **147**, 392 (1966).
W. Nolting and W. Borgiel, *Phys. Rev. B* **39**(10), 6962 (1989).
M. Ogata and H. Shiba, *Phys. Rev. B* **41**, 2326 (1990).
T. Oguchi, *Phys. Rev.* **117**, 117 (1960).
T. Oguchi, *Phys. Rev. B* **51**, 1385 (1995).
F. P. Onufrieva, *Zh. Eksp. Teor. Fiz.* **80**, 2372 (1981) [*Sov. Phys. JETP* (1981)].
S. G. Ovchinnikov, *Phase Transitions* **36**, 11 (1991).
S. G. Ovchinnikov and I. S. Sandalov, *Solid State Commun.* **47**, 367 (1983).
S. G. Ovchinnikov, *J. Phys. C, Solid State Phys.* **20**, 933 (1987).
S. G. Ovchinnikov and I. S. Sandalov, *Physica C.* **161**, 607 (1989).
S. G. Ovchinnikov and O. G. Petrakovsky, *J. Superconductivity* **4**, 437 (1991).
S. G. Ovchinnikov, *Zh. Eksp. Teor. Fiz.* **102**, 127 (1992a) [*Sov. Phys. JETP* **75**, 67 (1992)].
S. G. Ovchinnikov, *Zh. Eksp. Teor. Fiz.* **102**, 534 (1992b) [*JETP*, 75, (1992)].
S. G. Ovchinnikov, *Fiz. Tverd. Tela (Leningrad)* **36**, 2950 (1994) [*Sov. Phys. Solid State* (1994)].
S. G. Ovchinnikov, *Zh. Eksp. Teor. Fiz.* **107**, 796 (1995) [*JETP* **80**, 451 (1995)].
S. G. Ovchinnikov, V. K. Chernov, A. D. Balaev, N. B. Ivanova, B. P. Khrustalev and V. A. Levshin, *JETP Lett.* **64**, 642 (1995).

S. G. Ovchinnikov, *JETP Lett.* **64**(1), 25–31, (1996).
S. G. Ovchinnikov, *Usp. Fiz. Nauk.* **167**, 1043 (1997) [*Phys. Usp.* **40**, 993 (1997)].
S. G. Ovchinnikov, *Usp. Fiz. Nauk.* **169**, 869 (1999) [*Phys. Usp.* **42**, 779 (1999)].
A. Parolla and S. Sorella, *Phys. Rev. Lett.* **64**, 1831 (1990).
K. Penc, F. Mila and H. Shia, *Phys. Rev. Lett.* **75**, 894 (1995).
D. Pines, *Physica B.* **163**, 78 (1990).
N. M. Plakida and V. S. Oudovenko, *Phys. Rev. B* **59**, 11949 (1999).
N. M. Plakida, V. Yu. Yushankhai and I. V. Stasyuk, *Physica C.* **160**, 80 (1989).
A. N. Podmarkov and I. S. Sandalov, *Zh. Eksp. Teor. Fiz.* **86**, 1461 (1984) [*Sov. Phys. JETP* **59**, 856 (1984)].
M. Pompa, A. Bianconi, A. C. Castellano, S. Della Londa, A. M. Flank, P. Lagarde and D. Udson, *Physica C* **184**, 51 (1991).
P. Prelovsek, J. Jaklic and K. Bedell, *Phys. Rev. B* **60**, 40 (1999).
R. Preuss, A. Muramatsu, W. Von der Linden, P. Dieterich, F. F. Assaad and W. Hanke, *Phys. Rev. Lett.* **73**, 732 (1994).
P. H. P. Reinders, M. Springford, P. T. Coleridge, R. Boulet and D. Ravot, *Phys. Rev. Lett.* **57**, 1631 (1986).
J. J. Rhyne and N. C. Koon, *JMMM* **31–34**, 608 (1983).
T. M. Rice and H. J. Sigrist, *J. Phys. Condens. Matter.* **7**, L643 (1995).
J. L. Richard and V. Yu. Yushankhai, *Phys. Rev. B* **50**, 12927 (1994).
T. M. Riseman, P. J. Kealey, E. M. Forgan, A. P. Mackenzie, L. M. Galvin, A. V. Tyger, S. L. Lee, C. Ages, D. McK. Paul, C. M. Aegerter, C. R. Cubitt, Z. Q. Hao, T. Akima and Maeno Y. *Nature (London)* **396**, 242 (1998).
H. Romberg, M. A. Alexander, N. Nucker, P. Adelman and F. Fink, *Phys. Rev. B* **42**, 8768 (1990).
Yu. G. Rudoy and Yu. A. Tserkovnikov, *Teor. Mat. Fiz.* **25**, 196 (1975).
J. W. Rusul, *Phys. Rev. B* **39**, 663 (1989).
J. W. Rusul and P. Schlottmann, *Physica B* **163**, 689 (1990).
I. S. Sandalov and A. N. Podmarkov, "Povedenie anizotropnogo magnetica v sil'nom magnitnom pole (Behavior of Anisotropic Magnet in a Srtong Magnetic Field, Krasoyarsk", IF SO AN SSSR, 38 (1980) (Preprint SO AN SSSR, 156 F).
I. Sandalov, U. Lundin, O. Ericsson and B. Johansson, "Theory of strongly correlated systems. I. Intrsite Coulomb Interaction and the Approximation of Renormalized Fermions in Total Energy Calculations; II Including Correlation Effects into Electronic Structure Calculations", in *Physics of Strong Correlations in Electronic Structure and Model Calculation* by U. Lundin, (Acta Universitatis Upsaliensis, Uppsala 2000).
W. Schelp, W. Drewes, H.-G. Purwins and G. Eckold, *JMMM* **50**, 147 (1985).
C. M. Schineider, U. Pracht, W. Kuch, A. Chasse and J. Kirschner, *Phys. Rev. B* **54**, 15618 (1996).
J. Schmalian, M. Lander, S. Grabovski and K. H. Bennemann, *Phys. Rev. B* **54**, 4336 (1996).
J. Schmalian, D. Pains and B. Stoykovich, *Phys. Rev. B* **60**, 667 (1999).
L. Sham, *Phys. Rev.* **32**, 3876 (1985).

B. S. Shastry, *Phys. Rev. Lett.* **63**, 1288 (1989).
S. P. Shubin and S. V. Vonsovskii, *Proc. Roy. Soc. A* **145**, 159 (1934).
D. J. Singh, *Phys. Rev. B* **52**, 1358 (1995).
P. M. Slobodyan and I. V. Stasyuk, *Teor. Mat. Fiz.* **19**, 423 (1974).
R. Sollie and P. Schlottmann, *Phys. Rev. B* **41**, 8860 (1990).
J. Solyom, *Adv. Phys.* **28**, 201 (1979).
A. Svane and O. Gunnarsson, *Phys. Rev. Lett.* **65**, 1148 (1990).
L. Taillefer and G. G. Lonzarich, *Phys. Rev. Lett.* **60**, 1570 (1988).
H. Tasaki, *J. Phys. Condens. Matter.* **10**, 4353 (1998).
K. N. R. Taylor and M. J. Darby, *Physics of Rare-Earth Solids*, (Chapman & Nall, 1972).
T. Tohyama and S. Maekava, *Supercond. Sci Technol.* **13**, 17 (2000).
S. Tomonaga, *Progr. Theor. Phys.* **5**, 544 (1950).
S. V. Tyablikov, *Methods of Quantum Theory of Magntetism*, 2nd edn. (Nauka, Moscow, 1975).
V. G. Vaks, A. I. Larkin and S. A. Pikin, *Zh. Eks. Teor. Fiz.* **53**, 281 (1967a) [*Sov. Phys. JETP* **26**, 188 (1968)].
V. G. Vaks, A. I. Larkin and S. A. Pikin, *Zh. Eks. Teor. Fiz.* **53**, 1089 (1967b) [*Sov. Phys. JETP* **26**, 647 (1968)].
V. V. Val'kov and S. G. Ovchinnikov, *Teor. Mat. Fiz.* **50**, 466 (1982) [*Theor. Math. Phys.* **50**, 306 (1982)].
V. V. Val'kov and S. G. Ovchinnikov, *Zh. Eksp. Teor. Fiz.* **85**, 1666 (1983) [*Sov. Phys. JETP* **58**, 970 (1983)].
V. V. Val'kov and T. A. Val'kova, *Teor. Mat. Fiz.* **59**, 453 (1984).
V. V. Val'kov, T. A. Val'kova and S. G. Ovchinnikov, *Zh. Eksp. Teor. Fiz.* **88**, 550 (1985) [*Sov. Phys. JETP*, **61**, 323 (1985)].
V. V. Val'kov and T. A. Val'kova, *Fiz. Nizk. Temp.* **11**, 951 (1985) [*Sov. J. Low Temp. Phys.* **11**, 524 (1985)].
V. V. Val'kov, *Zh. Eks. Teor. Fiz.* **95**, 192 (1989) [*JETP* **68**, 109 (1989)].
V. V. Val'kov, *Teor. Mat. Fiz.* **76**, (1), 143 (1988).
V. V. Val'kov and D. M. Dzebisashvili, *Zh. Eks. Teor. Fiz.* **111**, 654 (1997) [*JETP* **84**, 360 (1997)].
V. V. Val'kov and D. M. Dzebisashvili, *Fiz. Tverd. Tela* **39**, 204 (1997) [*Phys. Solid. State* **39**, 179 (1997)].
V. V. Val'kov and D. M. Dzebisashvili, *The Physics of Metals and Metallography* **84**, 220 (1997).
V. V. Val'kov and D. M. Dzebisashvili, *Fiz. Tverd. Tela* **40**, 1674 (1998).
V.V. Val'kov, T. A. Val'kova, D. M. Dzebisashvili and S. G. Ovchinnikov, *Pis'ma v Zh. Eks. Teor. Fiz.* **75**, 450 (2002) [*JETP Lett.* **75**, 378 (2002)].
C. M. Varma, S. Schmitt-Rink and E. Abrahams, *Solid State Commum.* **62**, 10, 681 (1987).
A. V. Vedyaev and V. A. Ivanov, *Teor. Mat. Fiz.* **47**, 425 (1981) [in Russian].
A. V. Vedyaev and V. A. Ivanov, *Teor. Mat. Fiz.* **50**, 415 (1982) [in Russian].
A. V. Vedyaev and M. A. Nikolaev, *Zh. Eksp. Teor. Phys.* **82**, 1287 (1982) [*Sov. Phys. JETP* **55**, 749 (1982)].
A. V. Vedyaev and M. A. Nikolaev, *Teor. Mat. Fiz.* **59**, 293 (1984) [in Russian].

B. Velicky, S. Kirpatrick and H. Ehrenreich, *Phys. Rev.* **175**, 747 (1968)

S. V. Vonsovskii, *Magnetism* (Halsted, 1975).

S. V. Vonsovskii and Yu. A. Izyumov, *Usp. Fiz. Nauk.* **78**, 3 (1962).

S. V. Vonsovskii, Yu. A. Izyumov and E. Z. Kurmaev, *Superconductivity of the Transition Metals their Alloyes and Compounds* (Moscow, Nauka, 1977) [in Russian].

K. J. Von Szczepanski, P. Horsch, W. Stephan and M. Ziegler, *Phys. Rev. B* **41**, 2017 (1990).

A. Wasserman, M. Springford and A. C. Hewson, *J. Phys.: Condens. Matter* **1**, 2669 (1989).

A. Wasserman and M. Springford, *Adv. Phys.* **45**, 471 (1996).

B. O. Wells, Z. X. Shen, A. Matsuura, D. M. King, M. A. Kastner, M. Greven and R. J. Birgeneau, *Phys. Rev. Lett.* **74**, 964 (1995).

B. Westwanski and A. Pawlikovski, *Phys. Lett. A* **43**, 201 (1973).

C. N. Yang, *Phys. Rev. Lett.* **63**, 2144 (1989).

V. Yu. Yushankhay, G. M. Vujicic and R. B. Zakula, *Phys. Lett. A* **151**, 254 (1990).

V. Yu. Yushankhay, V. S. Oudovenko and R. Hayn, *Phys. Rev. B* **55**, 15562 (1997).

J. Zaanen, G. A. Sawatzky and J. W. Allen, *Phys. Rev. Lett.* **55**, 418 (1985).

R. O. Zaitzev, *Zh. Eks. Teor. Fiz.* **68**, 207 (1975) [*Sov. Phys. JETP* **41**, 100 (1975)].

R. O. Zaitzev, *Zh. Eks. Teor. Fiz.* **70**, 1100 (1976) [*Sov. Phys. JETP* **43**, 574 (1976)].

R. O. Zaitsev and V. A. Ivanov, *Fiz. Tverd. Tela. (Leningrad)*, 29, 1475 (1987) [*Sol. Phys. Solid State* **29**, 2554 (1987).

J. Zak, *Phys. Rev.* **23**, 2824 (1981).

F. C. Zhang and T. M. Rice, *Phys. Rev. B* **37**, 3759 (1988).

S. C. Zhang, *Phys. Rev. Lett.* **65**, 120 (1990).

A. K. Zvezdin, V. M. Matveev, A. A. Mukhin and A. I. Popov, *Rare-Earth Ions in Magnetically Ordered Crystals* (Nauka, 1985) [in Russian].

Index

Algebra of Hubbard operators, 15
Alternant lattice, 4
Angle resolved photoemission spectroscopy, ARPES, 23
Anisotropic ferromagnet, 129, 149
Anisotropic $s-f$ magnet, 150, 151
Annihilation operator, 1
ARPES data, 210
Atomic limit, 6
Atomic representation basis, 179
Averaging the diagonal X-operator, 86

Band limit, 5
Band structure, 210
Bogolubov (u, v) – transformation, 152
Bogolubov (u, v) – coefficients, 197
Bose-type quasiparticles, 41, 45

$CeCu_2Si_2$, 175
CeX, X = As, Bi, 188
Charge-transfer gap, 39
Chemical potential, 216
Commutation rule, 16
Cumulant, 87, 88, 89
Coulomb p–d interaction, 203
Coulomb p–p interaction, 203
Coulomb repulsion parameter, 2, 38
CuO_4-cluster, 196
CuO_6-cluster, 196

de Haas-van Alphen Effect, 175, 188
Diagram technique, 77
Dispersion equation, 154, 157
Dispersion relations, 184
d-type pairing, 225
Dynamical mean field approximation, 8, 12
Dyson equation, 123, 128, 129, 134, 141

Electron hole symmetry, 3
Exact diagonalization, 196
External operators, 90

Fermi-type quasiparticles, 41, 45
Fermi surface, 217
Ferromagnetic semiconductors, 176
Fourier transformation, 92

Generalized tight-binding method, 46
General rules for arbitrary order diagram, 112
Generating operator, 83, 95
Gor'kov equation, 142
Green's function, 20, 44, 50, 79, 184

Heisenberg model, 27
$HgCr_2Se_4$, 192
Hierarchy principle, 78
Hopping parameter, 2
Hubbard 1 decoupling, 49
Hubbard 1 solution, 6, 50

Hubbard model, 1
Hybridization, 34

In-gap state, 212, 213
Infinite dimension case, 12
Interaction line hierarchy, 97
Internal operators, 90
Irreducible diagram, 117
Irreducible part of the Green's function, 117
Irreducible operator, 222

Landau quantization, 189
Larkin Equation, 115, 118–120
Lehmann representation, 59
La_2CuO_4, 37
La_2NiO_4, 37
La_2MnO_4, 37
$La_{2-x}Sr_xCuO_4$, 216–218
Local density approximation, 56
Local eigenstates, 54
Localized magnetic moment, 27, 30
Longitudinal spin waves, 158
Lower Hubbard band, 39, 41
Luttinger liquid, 11
Luttinger theorem, 21

Magnetopolaron states, 189
Magnetopolaron effect, 176
Matsubara representation, 79
Matsubara time, 79
Mott-Hubbard gap, 52
Mott-Hubbard insulator, 20, 39
Multielectron model, 27
Multiorbital p–d model, 195, 200

Narrow band antiferromagnet, 171
Non-Fermi liquid effects, 191
$Nd_{2-x}Ce_xCuO_4$, 23

Ovals, 104
Overdoped metal, 210

p–d hopping, 38
p–d model, 38
Pairing of two operators, 84, 90

Pinning of chemical potential, 216
Periodic Anderson model, 33
Polarization operator, 160
Principle of topological continuity, 97
Propagator line, 92
Pseudo spin, 4
Pseudospin symmetry, 4
p-type pairing, 225
PuSb, 166

Quantum oscillations, 176

Rare-earth metals, 163
$ReAl_2$, 166
Root vectors, 17, 32, 78
Ruthenates, 218

s–$d(f)$ exchange, 30, 150
S-matrix, 81
S-pairing, 225
Self-Energy, 123, 127
Sign for arbitrary diagram, 99, 101
Single-ion anisotropy, 29, 71, 129, 158, 166
Singlet-triplet excitation, 199
Spectral density function, 8
Spectral weight, 46
Spin-flip transformation, 187
Spin polaron, 14
$Sr_2CuO_2Cl_2$, 210
Sr_2RuO_4, 218
Strength Operator, 123
$SU(3)$ group, 64
Sudden approximation, 23
Sum rule, 16, 46, 54

t–J model, 12
Temperature quantum oscillations, 191
The cell Hamiltonian, 207
Three-level forms, 69
Transition metal oxides, 37
Transformation laws, 66
Transverse interaction, 93
Transverse-longitudinal interaction, 98

Transverse spin waves, 158
Triplet pairing, 225
Two-level forms, 67
t–J–I model, 219

Undoped antiferromagnetic insulator, 210
Unitary operators, 64
Unitary transformation, 63, 178
Upper Hubbard band, 39, 41
UX, X=S, Se, Te, 166

van Hove singularity, 228
Vertex, 90

X-operator representation, 14, 25, 32, 35, 41

Wick's theorem, 77, 82, 85

Zhang–Rice singlet, 199